TEDBooks

Asteroid Hunters

CARRIE NUGENT

ILLUSTRATIONS BY MIKE LEMANSKI

TED Books
Simon & Schuster
New York   London   Toronto   Sydney   New Delhi

TEDBooks

Simon & Schuster, Inc.
1230 Avenue of the Americas
New York, NY 10020

TED, the TED logo, and TED Books are trademarks of
TED Conferences, LLC.

First TED Books hardcover edition March 2017

TED BOOKS and colophon are registered trademarks of
TED Conferences, LLC

SIMON & SCHUSTER and colophon are registered trademarks
of Simon & Schuster, Inc.

For information about special discounts for bulk purchases,
please contact Simon & Schuster Special Sales at 1-866-506-1949
or business@simonandschuster.com

For information on licensing the TED Talk that accompanies
this book, or other content partnerships with TED, please contact
TEDBooks@TED.com.

Interior design by: MGMT.design
Interior illustrations by: Mike Lemanski
Jacket design by: Chip Kidd

Manufactured in the United States of America

10 9 8 7 6 5 4 3 2 1

Library of Congress Cataloging-in-Publication Data has been applied for.

ISBN 978-1-5011-2008-4

ISBN 978-1-5011-2009-1 (ebook)

*To the next generation of scientists, who have so
much to discover*

CONTENTS

Asteroid Hunters

1 A Wild Frontier

I want you to imagine the solar system. I bet you're trying to recall an image from a childhood textbook: Mercury, Venus, Earth, Mars, Jupiter, Saturn, Uranus, Neptune. The planets are stately spheres, in a neat line leading away from the Sun.

Viewed this way, our cosmic neighborhood looks orderly and uncomplicated: a simple place that a child could draw. It seems thoroughly explored; after all, spacecraft have traveled to every planet. Each world has been measured and photographed. The solar system has been visited, explored, and mapped. Any remaining mystery must lie farther out, for future generations to discover.

I'm here to tell you that is not so. Our solar system is actually a wild frontier, teeming with different, diverse places: planets and moons, millions of objects of ice and rock. It is absolutely enormous; billions of kilometers (miles) across, so vast that we are only just beginning to figure out exactly where the solar system ends and interstellar space begins.

But even in our cosmic backyard, close to Earth, discoveries await us. I am talking about asteroids; the small, mysterious bodies that travel between the planets.

Every night, teams of scientists scour the sky for these objects. Every night, they discover new ones. It is a quiet effort, a steady accumulation of data over decades, a task that rewards coordinated teamwork. I am one of these asteroid hunters. It's a pretty cool job, and I love telling people about it.

When I tell people I'm a space scientist studying asteroids, they sometimes assume I'm a super-smart math whiz. The kind of person who skipped a bunch of grades and went to college when they were sixteen. Although I am good at math, school was difficult for me, and I didn't get straight As. But I was willing to work hard to satisfy my curiosity; I wanted to understand how things worked. As children, humans are naturally inquisitive about the world, and I was lucky to have encouraging parents. When I was five, my mom gave me a packet of litmus papers, chemical strips that change color when they are dipped in acids and bases. I ran around the house testing everything I could, trying to assemble a rainbow of results. So I was pretty young when I learned that science let you understand—and even better, predict!—the workings of the universe. By the time I got to college, I knew I wanted to study physics.

Becoming a scientist is a long journey, and at every step I found projects that were exciting, motivating me to continue. My path was not straightforward—when I began studying physics in college, I had no idea I would end up studying asteroids; in fact, I never took an astronomy class. But as it happened, my physics degree led me to study geophysics in graduate school, and that led me to study asteroids today. I love studying asteroids because they are relatively simple, just rocks in space. They can be understood with physics and described with elegant equations. For the most part, they are serene celestial bodies.

But for many people the word *asteroid* is synonymous with destruction; it brings to mind the extinction of the dinosaurs, or images from disaster movies of shattering buildings and cartwheeling cars. But large asteroid impacts are exceedingly rare. And as it turns out, there are actually things we can do now to lower the chance that someday one may harm us.

The idea that an asteroid impact can be prepared for, like one might prepare for a big winter storm, can come as a surprise. Metaphorically, asteroids seem to embody our lack of control over the universe. In literature, art, and popular culture, they are acts of God, cosmic phenomena that highlight our own powerlessness.

But the reality is quite different. As a species, we have the scientific understanding and technological prowess to actually do something about this particular problem. And it all starts with mapping the asteroids in our cosmic neighborhood.

Thanks to the hard work of generations of asteroid hunters, we have found almost all of the biggest, most hazardous objects. By the end of 2011, we had found over 90 percent of asteroids bigger than one kilometer across that get close to Earth; that is, those capable of massive destruction. And because the hunt for these objects has continued since then, that percentage is even higher today.

It is crucial we keep searching the skies. Not only would we like to find *all* the asteroids bigger than one kilometer across, it is also a good idea to find the slightly smaller, but still pretty big asteroids that are out there. Asteroid hunters are currently working toward a second target: finding 90 percent of the asteroids bigger than 140 meters across that get close to Earth. These objects are big enough to decimate a medium-sized country, and so far, only about thirty percent of these have been found.

Asteroid hunting is our responsibility to the rest of the planet. We are the only species able to understand calculus or build telescopes; the poor dinosaurs didn't stand a chance, but we do. If we found a hazardous asteroid with enough early warning, we could nudge it out of the way. Unlike earthquakes, hurricanes, or volcanic eruptions, asteroid impacts are a natural disaster that can be precisely predicted and, with enough time, entirely prevented.

As you will see, finding asteroids is a complex task that requires teamwork and patience. Asteroid hunters spend long nights on remote mountains, with only skunks and owls to keep them company. We use a robotic telescope that orbits Earth, diligently imaging the sky every eleven seconds. We send data to a centralized archive called the Minor Planet Center. And all this work is managed from a nondescript building in Washington, DC, by NASA's Planetary Defense Coordination Office, an unusual place that combines the futuristic work of defending Earth from asteroids with the bureaucratic reality of operating within the US government.

But let's back up a bit. What exactly is an asteroid?

Asteroids are commonly thought of as the rocky and metallic leftovers of the planet-building phase of our solar system. There are millions of them. The biggest are hundreds of kilometers across, while the smallest tracked ones are mere meters wide. Even smaller ones than that certainly exist, but are too tiny to see with today's telescopes. To the causal eye, asteroids are gray or brown; some are light in color, and some are so dark they look black. Most asteroids travel around the Sun alone, but some have asteroid moons. A few have two moons. So far, we've never seen one with three moons, but that doesn't mean we never will.

There are beautiful images of asteroids taken by spacecraft or radar imaging. Asteroids are generally lumpy. Like a Rorschach test, what you think an asteroid looks like often says more about you than the asteroid. Americans often default to "potato shaped"; in fact, asteroid 88705 is named Potato. After the Chinese spacecraft *Chang'e-2* took photographs of one asteroid, the mission scientists published a paper called "The Ginger-shaped Asteroid 4179 Toutatis." After the Japanese spacecraft *Hayabusa* imaged asteroid 25143 Itokawa, mission scientists compared its shape to a "sea otter," describing a distinct "head" and "body."

Where do asteroids live?

Most asteroids reside in the "main belt" between Mars and Jupiter, and never get close to Earth. Their orbits around the Sun haven't changed much over billions of years. Small in astronomical terms, they are quite large when viewed from a human perspective. One of the biggest is a main-belt object called Vesta, and it is 525 kilometers (326 miles) across. It's got roughly the same surface area as Pakistan.

Although it contains millions of asteroids, the main belt isn't at all the crowded place that you are probably picturing. I blame that misconception on that scene from *The Empire Strikes Back* where Han Solo plunges the *Millennium Falcon* into an "asteroid field" to evade the pursuing Empire. Princess Leia exclaims, "You're not actually going into an asteroid field?" Han Solo replies, "They'd be crazy to follow us, wouldn't they?"

Our heroes are dodging asteroids, which are flying in every direction, and the pursuing bad guys are being taken out one by one by collisions. C-3PO says, "Sir, the possibility of successfully navigating an asteroid field is approximately 3,720 to 1!" And Han, who is really selling these lines, says, "Never tell me the odds."

Clearly I love that scene.[1] But the real odds of safely navigating the asteroid belt in our solar system are a whole lot better.[2] It's pretty much 1:1. NASA has done this successfully many times, starting with *Pioneer 10* (which launched in 1972) to the Juno mission, which reached Jupiter in 2016.

Although there are millions of asteroids in the main belt, the fact is that each one is very small compared to the enormity of space. If you were to take all the known asteroids and squash them together to form a giant ball, that giant ball would still be smaller than our Moon. And the region in space where they orbit is so vast that if you were to stand on the surface of any one asteroid and look around, any other asteroids you could see would appear to be faint points of light.

The main belt isn't the only place in the solar system where you can find asteroids. There's a class of asteroids called trojans that hangs out along Jupiter's orbit, clustering a little before and a little after that planet as it orbits the Sun.

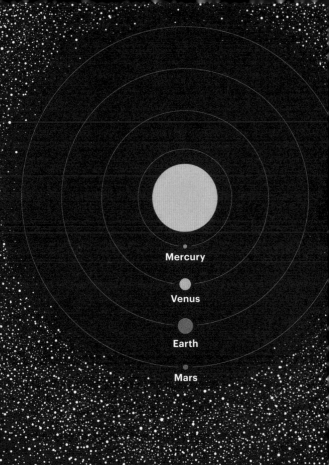

Mercury

Venus

Earth

Mars

Trojans

Main Asteroid Belt

Jupiter

Some of the moons of Jupiter and Saturn are probably asteroids that have gotten caught by the gravity of those giant planets. There are tens of thousands of rocky, icy objects beyond the orbit of Neptune in a region called the "Kuiper Belt," and perhaps many more beyond that in a region called the "Oort Cloud."

There are also comets. Comets, which occasionally light up the sky with their spectacular tails, have been known to humanity as long as the stars have. Traditionally, comets and asteroids were thought to be totally different types of objects—asteroids were made of rock or metal, and comets were made of rock and ice. As comets get close to the Sun, their ices—frozen CO_2 and water—sublimate, changing from a solid straight to a gas, leaving the surface. The departing gas brings dust with it, creating the beautiful tails we see.

New discoveries have blurred the line between comets and asteroids. Things we thought were asteroids have unexpectedly started to look like comets, and some asteroids may actually be "dead comets": comets whose ices have been burned off after many orbits around the Sun.

Several asteroids and comets have been visited by robotic explorers. As I write this, NASA's *Dawn* is orbiting Ceres. The European Space Agency's *Rosetta* orbits the comet Churyumov-Gerasimenko. The Japan Aerospace Exploration Agency launched *Hayabusa2* in 2014; it is on its way to the asteroid 162173 Ryugu to grab and return a sample of surface material. NASA has its own mission to an asteroid to grab and return surface material: *OSIRIS-REx* is slated to launch in September 2016 and will visit the asteroid 101955 Bennu.

Asteroids, and comets, offer an incredible opportunity to a space-faring species. With their diversity of orbits, many asteroids are much easier to get to than Mars but more difficult to get to than the Moon; stepping stones for astronauts as we venture ever farther from our home planet. Working on an asteroid would, of course, be technologically challenging. There would likely be large amounts of dust, and the barely-there gravity means that you can't walk around on the surface. As scientist Michael Busch described it to me[3], "It's not really landing on the surface. Because if you try to stand, the pressure from your legs is likely to throw you up off the surface again. And if you push too hard you fly off into space completely.... You can jump, and then spend two hours orbiting around and then land on the surface again. But it's not entirely clear what happens even when you touch the surface."

I am particularly interested in a class of asteroids called "near-Earth asteroids." These are often targets for spacecraft, and are often mentioned as potential targets for future crewed missions. *Near-Earth* is a broad term; it means only that the asteroid gets within 1.3 astronomical units of the Sun, where an astronomical unit is the distance between Earth and the Sun. So, many near-Earth asteroids spend most of their time out beyond the orbit of Mars, and plenty of near-Earth asteroids don't even get within 40 million kilometers (25 million miles) of Earth's orbit.

Astronomers are just beginning to map the population of near-Earth asteroids; quite simply, there are a lot of them. When I wrote this chapter, exactly 14,445 near-Earth asteroids had been discovered, but with over 100 being discovered every month, by the time you read this, that number is probably several thousand too small.

Finding asteroids is a challenging but solvable problem. Much like building a giant bridge, it's a problem that can be solved with math, hard work, and logic. It's not an easy task, but it's one well within our grasp.

2 Things that Hit the Earth

You might be surprised to learn that rocks from space hit Earth every day. Most of them are very small. In fact, the bigger the space rock, the less likely it is to hit Earth. Let's take a quick tour of a few notable impacts.

On any given day, about 90,000 kilograms (100 tons)[4] of dust and small rocks hit Earth. That seems like quite a lot to small creatures like us, but it's only 0.00000000000000000001 percent of Earth's mass. Put another way, that's less than one percent of the total weight of coffee consumed by humans each day. A scientist would say that's pretty negligible.

A couple of times a year, Earth passes through a region of space where a comet has previously been. The surface of the comet was warmed by the Sun, causing the water ice and frozen CO_2 on the surface to turn to gas, taking dust and tiny rocks with it. The comet left behind this cloud of tiny rocks, which gets pushed around a bit by sunlight and gravity. When the Earth passes through the cloud, those rocks burn up in our atmosphere, causing a beautiful display known as a meteor shower.

The best thing about meteor showers, in my opinion, is that they aren't fussy. They last several days, which allows you some flexibility in case the weather is poor. If you can find a really dark part of the world to watch from, you'll see more, but you can be in the middle of a big city and still see a few. I spotted a couple of the Geminid meteors one year from a backyard party in the middle of Los Angeles, which has pretty awful light pollution.

If you haven't seen a meteor shower, let me tell you, you are missing out. And to see one, you don't need a telescope or any fancy equipment. You just need a clear night at the right time of year, and you can easily find the dates of upcoming showers on the Internet. Grab a friend, a blanket, and some hot chocolate (or a beer, if that's your style), and lie on your back. Don't look at your phone—it will ruin your night vision. Look up, wait, and just shoot the breeze. Pretty soon you'll see flashes of light streak across the sky.

These streaks of light are caused by very small rocks; most are about as big as a grain of sand, some are as large as a pea. These rocks are traveling very fast relative to Earth, and the amount of energy (E) released when they hit our atmosphere is $E = 1/2\ mv^2$.

In this equation, m is the mass, and v is the velocity of the object. Even though your mass (m) is tiny, these grains are traveling at speeds of tens of thousands of kilometers per hour (miles per hour) relative to Earth, so the velocity (v) is large, and you get a fair amount of energy released; the equivalent of tens of thousands of lightbulbs lighting up all at once. That's bright enough for your eyes to see when you are standing on the ground. It's pretty amazing when you think about it; you're actually able to see the flash from something as small as a grain of sand from over 100 kilometers (60 miles) away.

Meteor showers are beautiful and harmless. But what happens when something a bit larger hits Earth?

Well, usually, nothing. Most of Earth is covered in water, so most of those objects end up in an ocean somewhere. And even with seven billion people on this planet, there's still huge regions of land that are uninhabited. But every once in a while one of these objects comes down in a populated area.

In November 1954, Mrs. Ann Hodges decided to take an after-lunch nap on her couch. She lived in a quiet town in the American South, about an hour's drive from Birmingham, Alabama. She lay down, covered herself with quilts, and went to sleep. What happened next was extremely improbable: a small rock, called a meteorite, crashed through her roof, bounced off her radio, and hit her on the side. Thanks to the roof and the blankets, she did not break any bones, but she was left with some spectacularly terrible bruises on her hip, hand, and arm.

The rock that hit Mrs. Hodges was once part of an asteroid, but when a rock from space travels through Earth's atmosphere and lands on the ground, we call it a meteorite. Meteorites are named after the locations where they are found, so this one was named Sylacauga after the small town near where Mrs. Hodges lived.

Although the meteorite itself did not kill Mrs. Hodges, it sparked an intense media frenzy and legal battle. Her landlord claimed ownership of the rock, since it had landed on her property. By the time Mrs. Hodges won the legal rights to the meteorite, interest had waned, and she couldn't find a buyer. Eventually she donated it to the Alabama Museum of Natural History. Some speculate that the stress of the legal battle and fame led to her divorce from her husband, and later her death from kidney failure at fifty-two, eighteen years later.

Take heart: it is extremely, extremely rare for a meteorite to hit a person. But because these things are difficult to verify in the historical record, it's hard to say exactly how rare it is. Mrs. Hodges is certainly the only American known to be hit. Every few months or so, I'll see stories on the Internet claiming someone was struck or killed by a meteorite; however, the facts of these cases rarely add up. In the hands of experts, meteorites are easily identified; oftentimes the suspect rock is just an unusual-looking Earth rock. And sometimes, explosions blamed on meteorites happen in areas with uncleared land mines from old conflicts. Sometimes it is easier to blame the cosmos for a death than to address a problem closer to home.

That's not to say that meteorites are entirely benign. After all, the Sylacauga meteorite, which hit Mrs. Hodges, would have caused her a lot more pain had it not first punched through her roof, clobbered her radio, and thumped into the quilts covering her. And that rock was only 3.9 kilograms (8.5 lbs) and about as big as a smallish melon. Let's look at something bigger.

On February 15, 2013, scientists were ready for an asteroid to zip close to Earth. A small asteroid (about 30 meters or 100 feet in size) known as 2012 DA_{14} (later named Duende) was going to get particularly close, zooming within the belt of geosynchronous satellites orbiting Earth. This object had been tracked for a year. Its orbit had been computed; there was no chance that it would hit our planet. The international scientific community planned observations. NASA organized a press conference with a panel of experts ready to answer any questions. In short, things were taken care of. There wasn't going to be any surprises.

But the solar system didn't care about these plans. That morning, something unexpected happened near the city of Chelyabinsk, Russia. A smaller asteroid, only about 20 meters (66 feet) across, sped through the atmosphere ten times faster than a fighter jet, coming in at a shallow angle. It was moving so fast that the molecules in the atmosphere didn't have time to get out of the way and built up in a layer of plasma that heated the rock. By the time it was about 38 kilometers (23 miles) above the surface, the heat and pressure was too much, and it burst apart in a series of explosions over four seconds, leaving behind debris and superheated air.

Residents saw a fireball speeding above, which briefly became brighter than the Sun, casting its own shadows. They felt the heat from the explosion on their faces, and some noted sunburns (though, being winter in Siberia, it was likely the residents were pretty pale). From the city of Chelyabinsk, this explosion was about 53 kilometers (33 miles) away, meaning it took more than two minutes between when they saw the flash of light and when they heard the explosion and felt the shock wave.

This event was captured by cell phone cameras, dashboard-mounted cameras in cars (popular in Russia), and surveillance cameras. This footage has been invaluable

Streaking faster than a fighter jet, a small asteroid exploded over the city of Chelyabinsk, Russia, in 2013. Causing injuries but no deaths, the explosion sent low-frequency sound waves across the globe.

scientifically, allowing scientists to measure the brightness of the explosions and the path the asteroid took. The images were shared immediately on the Internet; in fact, one scientist was alerted to the event by a tweet.

There's one YouTube video in particular that I think captures the feeling of what it was like to witness this event; a couple of men go outside to film the cloud trail when suddenly they are hit by the shock wave. There's a loud series of booms, the camera-lens appears to distort for a moment, car alarms go off, and windows shatter.

I don't speak Russian, but you can feel the shock and emotion in the men's voices as they react. I was so curious to know what the men were saying that when I learned that a summer student working with us at Caltech spoke Russian, I asked her to translate it for me. She did; it's mostly vivid and unprintable Russian expletives. But the parts that are printable are revealing; they talk of their "ears being squeezed" by the shock wave, and although one says "comet," another exclaims, "It's the war!"

A boulder-sized chunk of the original rock survived the journey through the atmosphere and crashed through the ice of Lake Chebarkul. Residents gathered at the lake, and one recalled in a *New York Times* article, " 'It was eerie,' Ms. Borchininova, a barmaid, said Saturday. 'So we stood there. And then somebody joked, "Now the green men will crawl out and say hello." ' "

Some residents took the event in stride. The *New York Times* also reported, "Out on the lake, an ice fisherman, who gave his name only as Dmitri, shrugged off the event. 'A meteor fell,' he said. 'So what? Who knows what can fall out of the sky? It didn't hit anybody. That is the important thing.' "

I agree, Dmitri, that is the important thing. But although no one was killed, hundreds were injured by shattering glass, and one man lost a finger. Having seen the flash in the sky, many people went to look out their windows at the cloud trail the object had left behind. Unfortunately, many of these people were then injured when the shock wave hit over two minutes later, shattering the windows in over 3,600 buildings. (So if you do happen to see a meteor flash in the sky, it's probably a good idea to find a safe place, away from windows, to hang out in for the next five minutes or so.)

The energy released by this explosion was so large it generated super-low-frequency sound, or "infrasound" waves that traveled the globe. The UN has 283 infrasound monitoring stations across the globe that listen for detonations of nuclear weapons. The Chelyabinsk explosion registered on twenty of these stations (including one in Antarctica) over three days as the shock waves traveled around the globe—two times.

But remember that other asteroid that was going to make a close approach that day? 2012 DA$_{14}$'s fifteen minutes of fame were stolen by this newcomer. Scientists quickly determined that the two events were unrelated; thanks to all the footage of the event, it was clear that the Chelyabinsk meteor had come from an entirely different direction than 2012 DA$_{14}$. Scientist Alan Fitzsimmons was quoted in another *New York Times* article, saying, "This is literally a cosmic coincidence, although a spectacular one."

It's worth noting that 2012 DA_{14} zipped by unremarkably; predictions of its trajectory were spot-on, and it didn't damage any of the satellites orbiting Earth. But why didn't we see the Chelyabinsk impactor coming? Well, for one, it was very small for an asteroid, only about 20 meters (66 feet), making it difficult to spot in the best of circumstances. Complicating matters, it approached Earth from the direction of the Sun, which is a blind spot for night-operating telescopes.

It's very lucky no one was killed by the Chelyabinsk meteor. Had the situation been just a little different, lives could have been lost. One lucky thing was that Chelyabinsk entered the atmosphere at a shallow angle, giving it more time to be heated by the atmosphere and distributing the force of the explosions over a broader area of ground. Additionally, the object was made out of relatively fragile rock, making it easier for the atmosphere to break it apart.

But not all asteroids are made of rock. A small fraction are made of nickel and iron. And as you can imagine, a metal asteroid packs more of a punch.

A nickel-iron asteroid hit what is now Arizona about fifty thousand years ago. There weren't any humans living in the area at the time to see it, but from a safe distance I'm sure it was spectacular. In fact, fifty thousand years later, it's still spectacular; it left behind a giant crater about 1.2 kilometers (0.7 mile) wide and 0.2 kilometers (0.1 mile) deep.

The asteroid that left this crater was 45 meters (150 feet) across. That's only two times as long as the one that hit Chelyabinsk; yet because this impactor was made of metal (and likely came in at a steeper angle), the damage was much more extensive. Even today, you can see fractures in rocks caused by the immense force of the explosion that excavated this enormous hole in the ground.

When it comes to asteroids, the objects that exploded over Chelyabinsk and Arizona were pretty small. In fact, they were smaller than almost all the known asteroids. The bigger the asteroid, the fewer of them there are, so the likelihood of an asteroid of a certain size hitting Earth decreases as you consider bigger and bigger asteroids.

Objects the size of the Chelyabinsk meteor hit Earth every hundred years or so. Every once in a very long time, something much bigger comes our way.

Certainly the most famous example is the asteroid that killed the dinosaurs 65.5 million years ago. That asteroid, or perhaps it was a comet, was about 10 kilometers across. It can be difficult to imagine just how enormous that is. Here's a trick that helps me. Ten kilometers is about 33,000 feet, which is roughly the cruising altitude of a modern airliner. So the next time you are in a plane, grab a window seat. Imagine a rock so enormous that, resting on the ground, it just grazes the plane's wingtip. It's so wide that it takes your plane a minute to fly past it. And it likely would have been moving over sixty times faster than an airplane when it hit.

The impact was almost unimaginably large. Scientists have used computer simulations to try to re-create the chain of devastating events that led to the extinction of over 75 percent of plant and animal species. But it happened so long ago that much of the evidence of the impact has been erased; in fact, it wasn't until 1980 when a father-son team of scientists, Luis and Walter Alvarez, recognized that this large impact had happened at all. But despite the tens of millions of years that have passed, two clues remain: an enormous crater just off the Yucatán Peninsula in Mexico, and a layer of iridium-rich sediment that can be found in locations across the globe. Iridium is an uncommon element to find on Earth but is prevalent in meteorites; it is the elemental fingerprint of the impacting asteroid.

I asked Rowan University paleontologist Kenneth Lacovara about this iridium layer, known as the Cretaceous–Paleogene boundary. There have been several suggestions that when the asteroid hit, the dinosaurs were already beginning to die out, plagued by a changing climate caused by volcanic eruptions. Given these new studies, did he really think it was an asteroid that caused this mass extinction?

"Absolutely," he said. "Below that layer you can find dinosaur bones. Above it, there's nothing."

It's pretty scary to think about something like that happening today. But I don't want you to lose any sleep over this. Consider that the last time this happened, humans didn't exist, and our closest ancestors were little mousy creatures. And, crucially, today we have computers and telescopes and teams of hardworking asteroid hunters searching the sky. Remember, we've already discovered over 90 percent of asteroids bigger than one kilometer across that get close to Earth, objects big enough to cause mass extinctions. And we asteroid hunters are persistent. Given the resources, it's only a matter of time until we find the rest.

3 Rules of Asteroid Hunting

The first rule of asteroid hunting is that you should never, under any circumstances, look at the Sun. The entire purpose of a telescope is to collect a lot of light and focus it into a very small place, either an eyepiece or an electronic detector that will record an image. Look too close to the Sun, and the telescope gathers a lot of the Sun's powerful radiation into one small point, melting the telescope's innards. Game over.[5]

Because of this, you can't hunt asteroids with telescopes on Earth during the day. Earth's atmosphere scatters light, brightening the sky even before the Sun rises above the horizon. Not to mention, asteroids are far too dim to be seen against the bright background of the daytime sky. So not only can you not look at the Sun, you can't look near it, either.

This means that most telescopes on Earth have a limited region of sky they can see. If you imagine Earth traveling around the Sun, there's always a line, called the terminator, which divides Earth into two halves: day and night. If you're looking for asteroids, and you can look only at night, you can see only things that are on the opposite side of the Earth from where the Sun is. You can see only half the sky at any given time. That means you can see only the part of the solar system that is currently farther away from the Sun than the Earth is, and beyond the terminator, so you can find only asteroids that are currently beyond Earth's orbit.

We build our lives around the cycle of sunrise, daytime, sunset, and nighttime. But when the Earth is viewed from space, you can see that when half the planet experiences day, the other half experiences night. Sunrise and sunset aren't distinct events; it's a constant process that sweeps across the planet as the Earth rotates. That sunrise-sunset line is called the terminator. Since astronomers using telescopes on Earth can only see asteroids at night, this means they can only find asteroids that are currently beyond Earth's orbit, leaving a blind spot between us and the Sun.

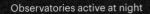
Observatories active at night

There are five main teams of scientists that hunt asteroids. These teams are called discovery surveys. Using telescopes on Earth are the Catalina Sky Survey in Arizona, Pan-STARRS in Hawaii, LINEAR in New Mexico, and Spacewatch, in Arizona. (The survey I work on, NEOWISE, uses a telescope out in space. More on NEOWISE later.) All four Earth-based surveys look at light in the sky in the visible wavelengths—the same wavelengths we see with our eyes. This means they're looking for sunlight reflected off the asteroids. So to more easily find asteroids, astronomers want to look when asteroids are the brightest—when they are reflecting the most light toward Earth. Like the Moon, asteroids are brightest (or "full") when the asteroid, Earth, and Sun are aligned. Outside this alignment, part of the asteroid will be in shadow, like a gibbous or crescent moon, and be less bright from Earth.

If your telescope is out in space, there's a few extra tricks you can use. The first is a sunshield. You have an instinctual understanding of sunshields—if you've ever tried to look toward the Sun to see something, you've held up your hand to block the majority of the sunlight. Holding up your hand to block light is a nice trick on Earth—it blocks the light overwhelming your eyes, while light scattered by the atmosphere and ground gives you enough illumination to see whatever it is you want to look at.

In space, you don't have any ground or atmosphere to scatter light, so when you block the Sun with a sunshield, things get even darker. And with a sunshield, you can look inside Earth's orbit. But still, you can never look at the Sun—there's always a "keep out" region or "exclusion zone" where you can't look.

And because space telescopes are extra sensitive, they also have to avoid looking at the Earth and Moon, as those are too bright as well.

The first rule of asteroid hunting—don't look at the Sun—has some surprising consequences. One might think that if you've got good telescopes and smart astronomers (which we do), you'd be able to find all the asteroids pretty quickly—in just a year or two. We look at the whole sky, we count what there is to see, tally them up, and maybe vacation on a beach somewhere, drinking fruity drinks with tiny umbrellas.

Unfortunately, it isn't quite that simple, and this is partly because, like a mob of small children, asteroids won't stay still while we count them. Imagine that asteroids are dozens of energetic schoolchildren running around the edges of a large grassy field. In the center of the field is a small group of teachers who want to count how many children are running around at this particularly crazy recess.

If the teachers can form a small circle, looking out, then together they can see the whole grassy field. If one teacher sees a girl in an orange dress running in front of him, he can count that child and tell the other teachers that he's counted the girl in the orange dress, and that she's running to the left. She's a known child, and nobody else should add her to the total number of children. The other teachers take note, and similarly share their observations. Through this process the small group of teachers could count all the children in fairly short order.

However, now imagine that there is a huge oak tree right next to the teachers, blocking their view of part of the field. The children are all running around at different speeds. Some are running quickly, so even if they are currently blocked by the tree, it won't take them very long before they run out into the part of the field the teachers can see and can be counted. But other children are moving slowly, picking flowers and looking at butterflies. Even if the teachers are very good at looking and counting, if the tree blocks their view, they are going to have to wait a very long time until the slow-moving children become visible and can be counted.

Now, unlike children, asteroids orbit in simple ways that can be described with elegant equations. However, also unlike children, asteroids aren't limited to a two-dimensional plane like a field—although many orbit the Sun in the same plane as Earth, some have orbits that swoop above and below that plane. But because the Sun creates a big region of space that we can't look at with telescopes, we're like the teachers with the oak tree—blind to one part of the field. Some asteroids have orbits that just so happen to keep them behind the Sun or difficult to see for years or decades. It's why Chelyabinsk wasn't detected until it hit the atmosphere. And that's why, even though there's been serious asteroid hunting for decades, we're still at it today.

The second rule of asteroid hunting is that you need to share your knowledge. So asteroid hunters have built a system to communicate their discoveries and coordinate their observations.

Say the astronomers at one observatory think they have found a new asteroid. How do they know, for sure, what they are seeing is an asteroid and not something else? When you are searching for the very faintest asteroids your telescope can see, sometimes artifacts of the telescope can masquerade as an asteroid moving across the sky. These artifacts can be a series of cosmic rays, or the edge of a flare from a bright star.

Well, if it is a real asteroid, it should have a predictable path. Artifacts like lens flares and cosmic rays won't move between the stars like an asteroid does. If it is an asteroid, we should be able to predict when it can be observed next, and by what size telescope. Any suitably equipped observer should be able to find it later.

Coordinating all this information from asteroid hunters around the world is a small group of people at the Minor Planet Center in Cambridge, Massachusetts. The Minor Planet Center has many functions; they archive all observations of comets and asteroids, and they are in charge of determining when a new asteroid has been discovered. Observers all over the world send them observations of known asteroids and possible new discoveries. They process about fifty thousand observations every day.

Most of their work is routine. But once in a while, something happens that emphasizes the importance of Rule #2—sharing your knowledge. Sometimes, you need to tell everyone what you've found because you urgently need everyone's help.

Tim Spahr was the director of the Minor Planet Center for nearly eight years. On October 6, 2008, he woke up and began his morning routine. Soy latte, check e-mail, wait for the caffeine to kick in. He had received an automated e-mail from the Minor Planet Center's computer about an asteroid Richard

Kowalski at the Catalina Sky Survey had discovered the night before. He told me this story several years later over (appropriately) a soy latte.[6]

"The computer is sort of gently suggesting I look at this orbit, because something was funny with it. And so I looked at the observations, and I recomputed the orbit, and I went to install it in our database, and it informed me that the close-approach distance was not a number."

Not a number (NaN) is an error you're familiar with if you work with computers. It's something you get if you, for example, try to divide a number by zero. Tim knows the Minor Planet Center's orbit-computing code inside and out. This error had never happened before, and the code was so well used it was surprising to find a new bug.

"And so I was like, *Hmm . . . not a number.* I don't really understand. So I did that calculation a different way, and it said the asteroid's distance from the center of the Earth in astronomical units was going to be 0.00002. So I had to get my calculator out and find that that was 3,000 kilometers from the center of the Earth, which is actually . . . well, the Earth is 6,300 kilometers in radius so that was a guaranteed impact at that point.

"Holy crap. I wasn't scared, because I knew it was a small object. It was more . . . whoa . . . that thing I thought was never going to happen in my career was going to happen . . . in twelve hours.

"I knew this was going to be a frantic, busy day, and the truth is I never made it into the office. There was literally no time for me to drive the forty-five minutes to the office. I stayed there the whole day, sitting at the computer, answering e-mails and processing observations."

And the observations were pouring in. Alerted to the situation, observers from across the globe were tracking the asteroid, which was given the temporary designation 2008 TC$_3$. They measured its position and brightness, and submitted those observations to the Minor Planet Center. There, Tim's colleague Gareth Williams was using them to refine his predictions of exactly where this small asteroid was going.

Gareth wasn't the only one calculating the asteroid's path; automated systems at NASA's Jet Propulsion Laboratory and the University of Pisa continuously check the list of known near-Earth asteroids for impact risks and post the results online. Scientists at those institutions were also working with the observers' data. Additionally, there is a small group of folks who calculate asteroid orbits as a hobby; one e-mailed Tim about the upcoming impact before the news had a chance to spread.

Even though it was daytime for Tim and half the planet, the network of asteroid hunters spans the globe, and observers for whom it was nighttime were able to spot the asteroid with their telescopes. Observations poured in from Australia, Russia, Slovenia, Switzerland, France, Italy, Spain, England, the Czech Republic, and Germany (just to name a few) . . . within seconds of each other. Anyone with a telescope big enough to see the asteroid was looking for it. As the Sun set across the globe, new observatories got a look; one of the last observations was by the John J. McCarthy Observatory, a volunteer-run telescope on the campus of a high school in New Milford, Connecticut.

"It was fantastic," Tim explained. "There was something like eight hundred observations of this object taken as it approached Earth."

Although deeply exciting for asteroid fans, this event was also a unique and valuable scientific opportunity. For centuries asteroids had been observed by telescopes, and meteorites had been collected on the ground. But never before did we have the chance to observe the same object as an asteroid in the sky and—hopefully—collect it as a meteorite on the ground. This was an enormous opportunity to test a range of theories linking asteroids and meteorites—and only twelve hours remained to study it in the sky. Careful observations were taken to see how the light reflecting off the asteroid changed over time, to measure its rotation. Special instruments were employed to carefully measure the color of the asteroid.

And the whole time, Tim was answering e-mails, and Gareth was incorporating the new observations into a prediction of when and where it was going to hit. As it happened, Steve Chesley and Paul Chodas of NASA's Jet Propulsion Laboratory were the first to predict the impact site: the Nubian Desert, in northern Sudan.

"We were able to predict the location of the impact so well that airplanes were alerted to the fact that there was going to be an impact in the area," Tim told me. "Two airplane pilots actually saw it explode in the atmosphere."

The pilots weren't the only ones. Despite exploding over a remote region, it was bright enough to be seen from far away. It happened to coincide with morning prayers and was seen by residents in northern Sudan and southern Egypt.

A desert is a pretty lucky place for an asteroid to hit if you want to collect meteorites. Much of the world is covered by water, and had it come down over an ocean, the meteorites would have been lost. And a dry area is ideal, as some asteroids are made of minerals that start to dissolve in water. But 2008 TC_3, only about 3 meters (10 feet) across, had exploded high in the atmosphere, 37 kilometers (23 miles) above the surface. Did any rocks survive, or had the entire object been vaporized?

Even if there was just a chance that some meteorites had survived, it was worth looking for them. SETI astronomer Peter Jenniskens teamed up with professor Muawia Shaddad and recruited students from the University of Khartoum to search for meteorites. At the location where the meteorites were predicted to be, the scientists and students stood side by side, arm's length apart, and walked across the desert. With the desert so large, and nothing like this done before, they didn't know if they were even looking in the right place.

After several hours of searching, however, success! Overall about 4 kg (9 lbs) of small meteorites were collected, a tiny fraction of the original asteroid's mass. They were named Almahata Sitta, which means "Station Six," after the remote train station that was closest to the impact. The meteorites themselves contained a fragile kind of rock never seen before in meteorites, and are the subject of current research.

But let's get back to Rule #2: Share your observations. The exciting story of 2008 TC_3, and all the discoveries it led to, was only possible because the Catalina Sky Survey, and all the other observers, openly and freely shared their observations. Scientists, engineers, and even dedicated hobbyists from across the world pitched in. Different teams computed trajectories and shared their solutions, providing checks of one another's work. Thanks to the Internet, and the Minor Planet Center, a worldwide campaign of observations was organized and executed in less than a day.

"It is a great way of operating," Tim said. "This is a scientific community, and science is done by verifying things and making everything available to anyone who is interested."

The discovery and impact of 2008 TC$_3$ was a huge, global event for asteroid hunters, and a great scientific accomplishment. But most people have never heard of it. That's because there was something else falling in October 2008.

"Coincidently, it was when the stock market was crashing," Tim said. "So it didn't get as much attention as it would have otherwise."

The first rule of asteroid hunting, "Don't look at the Sun," means it takes a while to find all the asteroids. The second rule, "Share your observations," is a consequence of this task being too large for any one person. I can think of other rules, such as "Be persistent," "Double-check your work," and, with a nod to *The Hitchhiker's Guide to the Galaxy* books, "Don't panic." But when it comes down to it, discovering asteroids takes patience, hard work, and a clear sky.

Although the vast majority of known asteroids have been discovered by professional astronomers working in teams, there's also a rich history of citizen scientists hunting for asteroids. These folks don't necessarily have astronomy degrees or funding from NASA—they just have a passion for space, a lot of patience, and a nice telescope. Like everyone else, they follow the asteroid hunting rules: they can't observe during the day, and they submit their observations to the Minor Planet Center. The rules of asteroid hunting are consequences of physics and our particular solar system—they apply to everyone equally, and have since the day the first asteroid was discovered.

4 The First Asteroid

In the late eighteenth century, German astronomer Johann Elert Bode thought he'd go down in history as a man who discovered a planet. Bode was living at a particularly interesting time for planetary science in Europe. Telescopes were bigger than ever. Kepler's equations describing the orbits of the planets around the Sun had won out over the Earth-centric models of old, and each planet's average distance from the Sun was measured. Here's a table of known planets at the time, along with their average distance from the Sun, measured in units of Earth-Sun distance (an astronomical unit, or AU).

Mercury	0.39
Venus	0.72
Earth	1.0
Mars	1.5
Jupiter	5.2
Saturn	9.5

Notice a pattern? Let's see how these numbers look on a graph (to your right).

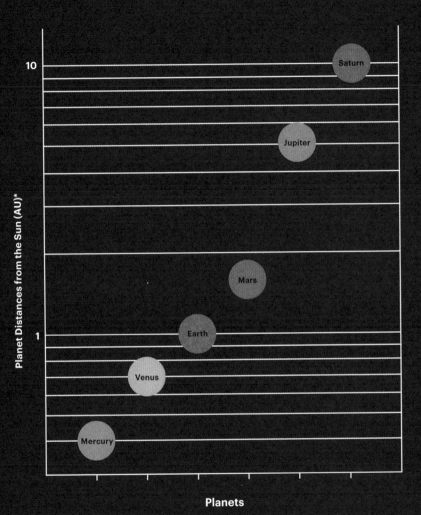

Planet Distances from the Sun (AU)*

10

1

Planets

*AU = Astronomical Unit.
Equal to 149.6 million kilometers, the mean distance
from the center of the earth to the center of the sun

The y-axis is logarithmic, which means the distance between the 0.1 line and the 1.0 line is the same as the distance between the 1.0 line and the 10 line. The planets are spaced out in a fairly even manner when plotted like this, except that it looks like there's something missing between Mars and Jupiter. Now, this pattern isn't particularly hard to notice, and although it's often credited to Johann Bode and another German astronomer, Johann Titius, various astronomers had been noticing the same thing since the early 1700s.

But this pattern had deep significance to Bode. A religious man, he believed that God must have arranged the planets deliberately, and this mathematical pattern was proof. A common belief at the time, this meant there must be something between Mars and Jupiter, and astronomers simply needed to find it. He wrote, "Can one believe that the Founder of the universe had left this space empty? Certainly not."

Bode used this pattern to predict the existence of a planet beyond Saturn and its distance from the Sun. When Uranus was discovered in 1781, it seemed a stunning validation. Despite the tradition of new astronomical objects being named by their discoverers, Bode leveraged his prediction to name the planet Uranus, after the Greek god of the sky. The discoverer of the planet, William Herschel, had wanted to name it *Georgium Sidus*, Latin for "George's star," after his patron King George III. (Personally, I wish "George" had won out; there's something comical about a planet named George. But "Uranus" is pretty funny in its own way.) After Uranus was discovered, Bode was even more convinced a planet must be hiding between Mars and Jupiter. And he was going to spear-head the search to find it.[7]

It became clear, however, that there was simply too much sky for this missing planet to hide in. On September 20, 1800, a group of six astronomers gathered to coordinate this search, led by Hungarian astronomer Baron Franz Xaver von Zach, director of the observatory in Gotha, Germany. Deciding that it was most logical to divide the sky into twenty-four sections, they needed to invite eighteen other astronomers to join their search. This elite team was to be known as the *Himmelspolizei*, or the Celestial Police. Invitations were to be sent to astronomers all across Europe.

However, it appears that not all invitations were sent. At least one astronomer, Giuseppe Piazzi, never received his. But that was all right, because he was already busy making a very precise map of the stars in the sky, using his fancy, custom-built telescope at Palermo Observatory in Italy. To make this star map, he observed the same stars three nights in a row. On January 1, 1801, he made his usual observations. On January 2, one star had moved slightly. Looking the third night, he saw it had moved again. He'd discovered something new.

From his letters, it seems that Piazzi knew he had found the new planet Bode was seeking. I imagine he must have been ecstatic. But the unfailingly cautious Piazzi was careful with his language when he reported his discovery, even when he was writing to his trusted friend, fellow Italian astronomer Barnaba Oriani:

"I have announced this star as a comet, but since it shows no nebulosity, and moreover, since it had a slow and rather uniform motion, I surmise that it could be something better than a comet. However, I would not by any means advance publicly this conjecture. As soon as I shall have a larger number of observations, I will try to compute its [orbit]."

We now know what Piazzi saw was not the new planet astronomers were hoping to find. But he was right to surmise it was something better than a comet. He had discovered the first asteroid.

But until he was entirely certain of his discovery, and had the orbit to support his claims, Piazzi shied away from announcing his discovery to the world. He did decide to take one risk, which was to write Bode: "On January 1st I discovered a comet in Taurus ... Please, let me know if it has already been observed by other astronomers, for in this case I will not bother with the calculation of its orbit."

Reading between the lines, I wonder if Piazzi wasn't daunted by the orbit calculations, and was in some part hoping that someone else had beat him to the discovery, relieving him of that responsibility.

Bode, in stark contrast to Piazzi, couldn't contain his excitement. Unfettered by any lack of confidence, he took it upon himself to announce to the world that a new planet had been found. Astoundingly, he even named the object "Juno." (His tactic of preemptive naming had proven successful with Uranus, so why not try again?) Baron Franz Xaver von Zach evidently also felt he had ownership of this discovery, as the leader of the Celestial

Police (never mind the fact that Piazzi's invitation hadn't been sent), and started to call it "Hera."

Piazzi had, as was his right, named it himself. He chose *Ceres Ferdinandea*, to honor the patron goddess of Sicily, and his patron, King Ferdinand. Piazzi's friend Oriani had to break the bad news to him that the astronomers in Germany were calling it something else. Piazzi wrote back, "If the Germans think they have the right to name somebody else's discoveries they can call my new star the way they like: as for me I will always keep it the name of *Cerere* [Ceres] and I will be very obliged if you and your colleagues will do the same."

I think this is an extraordinarily even-tempered response. If I thought I had found a new planet and someone else not only announced it publicly but also tried to name it, my response might be unprintable in a family-friendly book like this one.

(Eventually the Germans did come around, though not without some complaint. Von Zach also wrote Oriani, who seems to have been mediating the whole affair. "I shall continue to call it Ceres, but I beg Piazzi to drop Ferdinandea because it is a bit too long.")

The astronomical world was fired up, and everyone wanted a copy of Piazzi's observations so they could observe it themselves. But poor Piazzi was having a rough time of it. He had faithfully observed his new object every night he could. From one night to the next, the object was easy to follow in the sky; it was moving in the same direction, and it didn't move much night to night. But over the course of several weeks, Ceres and Earth were moving relative to each other as they orbited the

Sun. This caused Ceres to appear in the sky earlier and earlier each night. Inevitably, it started to appear too close to sunset, when there was too much light in the sky to see the faint object. Piazzi was coming up against Rule #1: Don't look at the Sun.

Piazzi knew he had reached a crucial point. If his observations were good enough, then, theoretically, he could calculate an orbit from them and predict exactly where in the sky this object would appear again. But, perhaps exhausted from his efforts, he became very ill as soon as his discovery moved out of view.

After he had recovered, Piazzi tried valiantly to calculate an orbit for his discovery, but the calculations just weren't working, and the math was challenging for him. He repeatedly wrote his friend Oriani for help, asking for more equations and explanations. He desperately wanted to calculate the orbit himself before releasing his observations to the world, as the orbit would confirm the discovery, adding weight to his claim. But pressure from the other astronomers was mounting (Rule #2: "Share your observations"). Everyone was itching to see this new planet for themselves.

Not everyone was being nice about it. Britain's Astronomer Royal, Nevil Maskelyne, has been remembered by history as a bit of a jerk. (If you have read Dava Sobel's *Longitude*, you may remember Maskelyne as John Harrison's nemesis). At this time, he wrote a particularly nasty passage:

"There is great astronomical news: Mr. Piazzi, Astronomer to the King of the two Sicilies, at Palermo, discovered a new planet the beginning of this year, and was so covetous as to keep this delicious morsel to himself for six weeks; when he was punished for his illiberality by a fit of sickness . . ."

Piazzi, unable to calculate an orbit, eventually caved and shared his observations, which were quickly distributed. Others tried to fit orbits as well, but similarly met with frustration. Piazzi must have been relieved when he learned he was not the only one who could not solve for an orbit. Months after it disappeared into the evening sky, Ceres could not be found again, despite searches by astronomers all across Europe. The frustrated Baron von Zach started to complain to Piazzi's friend Oriani: "There are some astronomers who are starting to doubt the real existence of such a star. . . . Now I cannot conceive, how an observer as experienced as Piazzi . . . could incur such mistakes in his meridian observations."

This public doubt must have felt like confirmation of Piazzi's worst fears, that his observations were somehow flawed. As it turns out, it wasn't Piazzi's fault that Ceres's orbit could not be predicted. Existing methods of the time used the assumption that the planets traveled around the Sun in circular paths, when in reality they traveled on a specific geometrical path called an ellipse. These particular ellipses were generally almost circular, however, so the circular approximation was usually pretty reliable, yet challenging enough to solve in a time without computers or calculators. However, this method, and other existing methods, were simply not good enough to predict where Ceres was going, and prevented astronomers from finding it again.

Luckily for everyone, an honest-to-goodness mathematical genius was alive, and he thought the problem looked interesting. Carl Friedrich Gauss, then twenty-four, turned his attention toward predicting where Ceres could be observed next. Accounting for the elliptical path and the movement of the Earth and Ceres relative to the Sun, Gauss formulated a complicated equation that he then solved using various approximation methods, some of which he invented on the spot (including the fast Fourier transform. If you know what that is, you're already impressed. If you don't, suffice it to say it was named one of the "Top 10 Algorithms of the 20th Century" by the IEEE journal *Computing in Science & Engineering*, due to its "influence on the development and practice of science and engineering in the 20th century."). Based on this new math and Piazzi's observations (which were, in fact, excellent), Gauss predicted where Ceres should be in the sky. Baron von Zach, using these new predictions, observed Ceres on December 7, 1801, almost a full year after its discovery—though he didn't know it at first.

Ceres, like all asteroids, looks just like a star when viewed through a telescope. The best way to distinguish between a star and an asteroid hasn't changed much in the last two hundred years—you look at the same patch of sky once, and then again a little later, and see if what you think is an asteroid has moved relative to the stars behind it. Because of the limitations of the telescopes of the time, this meant von Zach needed to observe the same patch of sky over the course of two nights.

But the weather wasn't on von Zach's side. It was cloudy for weeks, and he couldn't get the second observation he needed. Cooped up and irritated, he again complained to Oriani in a December 18 letter:

"What is going on with the *Ceres Ferdinandea*? Nothing has been found as yet either in France or in Germany. People are starting to doubt: Already skeptics are making jokes about it. What is Devil Piazzi doing?"

However, on December 31, the skies above were finally clear for von Zach, and he was able to get the second observation of the sky that confirmed the rediscovery of Ceres. Another German astronomer, Heinrich Wilhelm Matthias Olbers, found it independently two days later. Thanks to Gauss's excellent mathematical solution, Ceres was found, and Piazzi's good name was upheld.

It was starting to become clear, however, that something wasn't quite right about this new "planet." It wasn't quite as bright as it should be, meaning it was much smaller than the other planets. And three months after finding Ceres again in the sky, Olbers discovered another "planet" between Mars and Jupiter, which was named Pallas.

William Herschel (who discovered Uranus) started to suspect that Ceres and Pallas weren't planets, or comets, and were instead something new. He put forward the name *asteroid*, from the Greek for "starlike," because, aside from their motion, they looked exactly like stars through a telescope. This didn't please Piazzi, who thought it demoted his discovery. But as time went on, several more asteroids were discovered, strengthening Herschel's theory that they were a new category entirely.

With this new classification, it would seem that Bode, von Zach, and the Celestial Police never found their planet between Mars and Jupiter. And it's easy to understand why Piazzi thought calling Ceres an asteroid instead of a planet lessened its significance; he couldn't have predicted just how numerous asteroids were, or how important they would become scientifically.

But as it turns out, Ceres *is* special. It is the largest asteroid in the inner solar system, over 900 km (560 miles) across. As I write this, a NASA spacecraft, *Dawn*, orbits it, taking photographs and scientific measurements. And in fact, although many call it an asteroid, it is also considered a "dwarf planet," since it has planetlike features, such as enough gravity to mold its shape into a sphere.

And this brings up an interesting point: classification, the human practice of sorting things into categories, is sometimes arbitrary and can have unintended consequences. Calling Ceres an asteroid instead of a planet made it seem less important to many astronomers, and perhaps less worthy of study. Today, with the images of its surface from *Dawn*, we know it is a unique world with fascinating geology and chemistry, not just an object orbiting the Sun but a place one could visit.

But no matter how you classify it, the discovery of Ceres was a big deal. It was the first asteroid, the first indication that there might be more to the solar system than just planets, moons, and comets. And despite the advances in technology, and the two hundred years that have passed, this story of discovery is very familiar to asteroid hunters today. Much remains the same: the need for multiple observations of asteroids to determine their orbits, the challenges of losing objects to the evening sky, and the international collaboration needed to track these objects over time.

5 Terrestrial Asteroid Hunting

In the early twentieth century, technological advancements gave rise to bigger and better telescopes. Now it seemed that you couldn't point your telescope at the sky without discovering a new asteroid. Surprisingly, these new discoveries were no longer a thrill; having made a discovery, astronomers felt obligated to also calculate the new asteroid's orbit. And even with the mathematical process well documented (thanks to Gauss), it was quite a lot of math and a bit of a pain. Worse, it was a distraction from other things astronomers wanted to study.

Astronomer Edmund Weiss was so irritated that he started calling asteroids "those vermin of the sky." Joel Hastings Metcalf, an American astronomer and minister, was only slightly more polite. In 1912 he said, "Formerly, the discovery of a new member of the solar system was applauded as a contribution to knowledge. Lately it has been considered almost a crime."

The two world wars put a hold on asteroid research and many other forms of astronomy, but by the 1960s there was a resurgence of interest. Astronomer Charles Kowal recalled, "Young astronomers wanted to learn about the asteroids (perhaps to the chagrin of their professors). . . . Most of all, they learned that the asteroids were quite exciting!"

With fresh eyes, a new generation of scientists saw potential, not nuisance, in these small bodies. Teams of scientists banded together to start the first dedicated surveys to hunt these objects. One of the earliest modern asteroid surveys was the Palomar–Leiden survey, operating from 1960 to 1977. This used two telescopes, Palomar Observatory in California, and the Leiden Observatory in the Netherlands. Led by a husband-and-wife team in the Netherlands, Cornelis and Ingrid van Houten, and Tom Gehrels at Palomar, they discovered 4,620 asteroids and comets. Around the same time, another husband-and-wife team, Nikolai and Lyudmila Chernykh, discovered hundreds of asteroids from the Crimean Astrophysical Observatory in Ukraine. In 1973 the Palomar Planet-Crossing Asteroid Survey was founded by Eleanor "Glo" Helin and Eugene Shoemaker.

These surveys used glass plates, or film, to image the sky. There's not a whole lot of people who remember what it was like to find asteroids this way. One of these people is Tim Spahr, who you may remember from Chapter 3 ("Rules of Asteroid Hunting"). He got his start hunting asteroids as a student at the University of Arizona. Together with fellow student Carl Hergenrother and faculty sponsor Stephen M. Larson, they ran the Bigelow Sky Survey in the early 1990s. One morning, Tim took some time to tell me about the process.

Tim and Carl would work with special film from Kodak, known as T-MAX 400. This was a specialty film, prized by astronomers for its sensitivity. To find the most asteroids, you wanted to take as many photos of the sky as you could before the Sun rose, so the shorter your film's exposure time, the better. With Kodak T-MAX 400, Tim and Carl could use exposure times of six to ten minutes.

The film itself had to be cut to size in a darkroom and put into a special light-tight box on the telescope. The telescope had to be moved so the film could be loaded; even though it was on a pivot, the telescope was a heavy contraption of glass and metal. Tim compared moving the telescope to pushing a Volkswagen Rabbit (a small car) in neutral gear. Once the film was loaded, Tim or Carl would push (the proper word here is *slew*) the telescope back toward the sky.

Then they had to point the telescope at exactly the right part of the sky. Today, robotic controls allow telescopes to point precisely via computer commands. But back then Tim and Carl would use "guide stars," particularly bright stars, to find their way to the small patch of sky they wanted to see. They did this so many times, Tim told me, that "Carl and I got this bizzaro thing where we were actually memorizing the guide stars." They did this so many times they no longer needed charts; they knew the stars by heart.

Once they found the right patch of sky, they would secure the telescope and lift the cover off the film. But this wasn't the time to take a break. As Earth turns, the stars appear to move slowly across the sky. So during an exposure, the telescope would automatically move, tracking with the stars. But the telescope's tracking system wasn't perfect, so to get really crisp images Carl or Tim would use a small guiding telescope, attached to the big one, to track the guide stars during the exposure. By slightly adjusting the big telescope so that a particular star was always in the middle of the guide telescope's illuminated crosshairs, they could get a nice, sharp image. But,

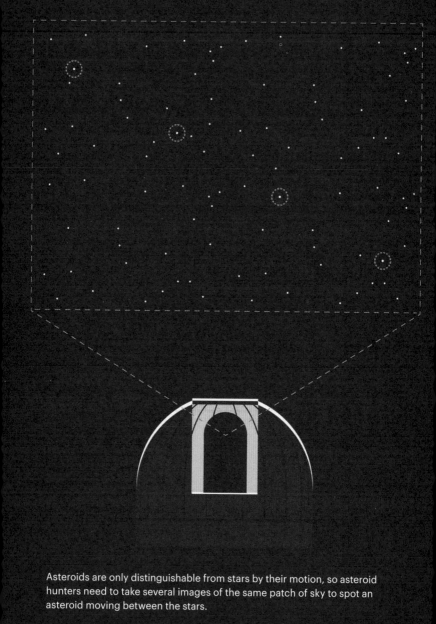

Asteroids are only distinguishable from stars by their motion, so asteroid hunters need to take several images of the same patch of sky to spot an asteroid moving between the stars.

From far away, asteroids look like points of light. But up close, each one has a unique shape and surface.

Tracking an asteroid over several nights lets astronomers precisely measure its orbit around the sun.

of course, this task, like the others, wasn't easy. Tim recalled that there was a loose wire to the circuit that illuminated the crosshairs, and every once in a while they'd put their eye to the eyepiece and get a nasty shock to their brow ridge. He added, "I also froze to it one time."

After the six- to ten-minute exposure they would cover the film, slew the telescope back, and remove the film, replacing it with a fresh piece. This was then repeated, covering more patches of sky, and getting duplicate images a little while later to see if anything had moved.

Carl or Tim would alternate observing and developing the film; they developed the film themselves, bathing it in a series of chemicals. And then, as soon as they could, they would load the developed film into a specialized piece of equipment called a stereo microscope to find the asteroids.

The stereo microscope exploits the way human brains interpret what our eyes see. The microscope would hold two pieces of film of the same part of the sky taken at different times. Stars, which don't move relative to each other, would appear in both images in the exact same plane. But asteroids, which do move, would appear to the viewer as floating a little in front of, or a little behind, the flat plane of stars.

Spotting an asteroid this way, Tim explained, "provides an adrenaline rush that cannot be met any other way in your life." Although it's fun to discover a new asteroid independent of the circumstances, there's something different about this method, when the asteroid literally appears to be jumping out at you, wanting to be found. To this day, Tim says, he'll occasionally feel a thrill when two streetlights line up just right, echoing the asteroids he used to seek.

But searching the film for asteroids was a slow process. Each small portion had to be searched methodically. By the end of several weeks, Tim told me, "We were exhausted, sleep-deprived, and grumpy. It was the most fun we had in our entire lives."[8]

At the same time, a new technology was becoming more widely used: CCDs, or charge-coupled devices, are the same technology that makes your digital camera work. CCDs offered much promise for asteroid hunting. They were more sensitive than film, meaning that you could spend less time taking an exposure and get more images each night. They didn't require chemical baths to develop, again saving time. And, since the images were digital, they could be scanned by computers for asteroids, eliminating the thrilling but grueling step of examining film, inch by inch, in a stereo microscope.

The path to CCDs was not an easy one. Bob McMillan, current head of the Spacewatch survey, was a young scientist on the project when it was founded by Tom Gehrels in 1980. Spacewatch was the first asteroid survey to use CCDs. These early CCDs were tiny and covered much less of the sky than film could. And they didn't come in the nice packages they do today—they were just delivered as a bare chip. Everything else— the mount for the chip, the connections to the computer, the software to interpret the output—had to be created from scratch.

Bob recalled the do-it-yourself nature of those days.

"When we first assigned our engineer to develop a CCD camera, he visited the Jet Propulsion Laboratory, where Jim Westphal had been doing the same in his lab. Jim was using those hemispherical stainless-steel spaghetti bowls as vacuum housings to cool the electronics. Our engineer saw that and for some reason thought that those bowls would be too expensive, so instead for our camera he used a half-gallon size tin coffee can (which is also used for vacuum packing). We still have that camera on a shelf in our lab. It was cooled with dry ice, very messily."

During the 1980s CCD technology matured. Bigger chips became available, allowing larger portions of the sky to be seen. As mounts and software to read them became more standard, others started to follow the path blazed by Spacewatch. By the early '90s CCDs became so efficient that they outperformed film surveys. Tim and Carl were among the last astronomers to hunt asteroids using film.

One of the first big CCD surveys was the Near-Earth Asteroid Tracking (NEAT) program, led by Glo Helin. Whereas Tim and Carl, working like crazy, discovered dozens of asteroids, NEAT quickly started finding hundreds.

Glo founded NEAT, in part, because she was concerned with the potential hazard from asteroids. In the online archive Web of Stories, she referred to them as "little marauders that come in and occasionally smack us, if you will."

As she and her team discovered more and more asteroids, they were reminded of the possibility that one could hit Earth. When NEAT first discovered an asteroid, there was generally a large amount of uncertainty as to where exactly it was going. She recalled, "All my colleagues emphasize that we'll have plenty of time to prepare for [a hazardous asteroid]. As an observer, I've been on the spot when we have had early computations of orbits that are on a collision course with the Earth. I guess I've been feeling somewhat closer to that possibility, having to ride out the tense moments or sometimes days until we get more data."

As it turned out, every asteroid NEAT found is on an orbit that keeps it away from our planet. Of course, Glo couldn't have known that at the time, and those uneasy times of uncertainty before a collision was ruled out motivated Glo to keep searching.

Tom Gehrels, who founded the Spacewatch survey, appears to have felt similarly. In a 1984 *People* magazine interview, when asked about the possibility of an asteroid headed on a collision course with Earth, he said, "That damned thing may be out there right now."[9]

Today, asteroid-hunting surveys all use CCD technology. NEAT is no longer operating, but Spacewatch is. Additionally there is the Catalina Sky Survey, Pan-STARRS, LINEAR, and NEOWISE, which surveys from Earth orbit. You may have noticed that all these surveys are American. Although there is a vibrant international community of scientists studying and observing asteroids, over 95 percent of near-Earth asteroids have been discovered by NASA-funded surveys.

I had a conversation one morning with the director of the Catalina Sky Survey, Eric Christensen, and he told me how his team hunts for asteroids.

Every night, when the sky is clear, two Catalina Sky Survey observers drive up to two separate telescopes in the Santa Catalina Mountains (the survey's namesake) outside of Tucson, Arizona. As they drive, gaining altitude, the landscape changes from saguaro cactus and desert to ponderosa pine forest. The temperature drops about 17°C (30°F). The change in environment is so dramatic that biologists have termed these mountains "sky islands"—forest ecosystems isolated from each other by the dry, hot desert between.

"It's one of my favorite parts of the job," Eric told me. "I always love getting up to the mountains." He said it's especially nice in the summer, when observing means an escape from the 43°C (110°F) heat. And since the Catalina Sky Survey consists of only about ten people, half of whom are astronomers, even the director gets to observe occasionally.

There are two sites: Catalina Station and Mt. Lemmon Station. The observatory at the Catalina Station was established in 1963 by the Lunar and Planetary Laboratory at the University of Arizona; it was built for a variety of astronomical purposes, not just to find asteroids. The Mt. Lemmon Station was originally part of a network of air force radar monitoring bases. Operating from 1956 to 1970, it was home to hundreds of people, and had three radar towers, an electrical generation plant, barracks, a mess hall, and even a two-lane bowling alley.

It's much quieter today. Some of the original buildings remain and have been turned into classrooms, dorms for observers to sleep in during the day, and a home for the survey's resident caretaker.

Eric describes the facilities as having a "Cold War feel."

"The buildings are painted a drab green, the equipment dates from that era and earlier eras. The telescopes in use are older, built in the '70s, and have been refurbished for their current task. It's not the most modern . . . it's utilitarian. But it's a beautiful site, and the views are commanding."

At night, it is very quiet, except for the mechanical creaking of the moving telescopes as they pivot to see different parts of the sky. Observers work alone in a heated control room. Up in the mountains, the quiet and isolation from humanity sometimes makes observing feel more like a camping trip than being a space explorer.

"I've personally seen bobcats, wild turkeys, deer, skunks, bats, and even a black bear once, a few hundred yards from the observatory," Eric told me. Other observers have seen mountain lions, and one was surprised to discover a skunk keeping him company in the control room.

"The observer, remarkably, managed to stay calm and shoo it out into the vestibule, assuming it would wander out the open door. But a few minutes later he heard a rustling behind him, as the skunk had found another way into the control room. The quality of construction is somewhat 'rustic,' so it probably found a hole in one of the wall panels." The observer decided to call it a night and left the site. Surveying at that site was paused; it took a few days for the skunk to be live-trapped and relocated.

Catalina's astronomers use their years of experience to choose where the telescope is going to look that particular night. Catalina's telescopes are good for finding asteroids because

they can see large patches of sky at once; their field of view is much wider than most.

The typical observing pattern goes like this: Point at a patch of sky, gather light for thirty seconds to create an image. Point the telescope elsewhere, and take more images. Then point back to the first patch of sky about fifteen minutes later and take another image. Repeat until you've got four images of the same part of the sky, separated in time.

Eric described this as a "low-resolution movie," with four frames over the course of an hour. If there's an asteroid in those images, it will appear to be traveling between the stars, noticeable due to its movement. The survey uses software to scan the images, looking for these objects. But the whole time, the observer is there in the forest, checking the software's output and looking at the images.

"You're making adjustments and solving problems, it's a busy and interesting night," Eric said. "And there's always the instant gratification of asteroid observing that you don't get from other types of observing. A few minutes after you take the data, you can be looking at a potential near-Earth asteroid. When you make a discovery, you're the first person to see the photons from that individual asteroid."

Then, Eric explained, the chase is on. After a new discovery is made, observations over several nights are needed so that the asteroid's path can be computed accurately, years into the future. Astronomers at the Catalina Sky Survey "follow up" their discoveries, keeping an eye on where they're going over the course of several nights, until the orbits can be determined precisely. An orbit is a unique identifier of each asteroid, and computing the orbit lets you know if that asteroid has been seen before or not.

It isn't unusual to discover even more new asteroids in the follow-up images. "We've found near-Earth asteroids when following up comets, comets when following up a near-Earth asteroid. Space is swimming with all these unknown objects. I don't tend to be too romantic about it, but you find objects where you look for them. And if you don't look, you're not going to find. Being open to surprising discoveries has really paid off."

On a good night, an observer can find five to ten new near-Earth asteroids; new camera updates may soon boost that number to thirty. And in addition to finding hundreds of near-Earth asteroids each year, one of Catalina Sky Survey's claims to fame is that they are the only survey that discovered two relatively tiny asteroids right before they later hit Earth, including 2008 TC$_3$, which as you'll recall exploded over northern Sudan.

The hazard that near-Earth asteroids can pose is something Eric does not forget. "With asteroids, there may be something out there that could directly affect our lives or our civilization," Eric said. "But I don't tend to dwell on it—I see it as a small statistical risk rather than an immediate personal risk."

This is the attitude of most asteroid hunters today—we know finding asteroids is important work that needs to be done. We are the modern-day Celestial Police. But thanks to the work of previous astronomers, we've got a better idea of what's out there, and the risk looks a little smaller than it did thirty years ago. We also know that, statistically, asteroids are not in the top ten of likely causes of death. We're far more likely to die from heart disease, a car accident, or even the flu.

It's the practical nature of finding asteroids that appeals to many asteroid hunters, including me. Asteroids hold a unique position in astronomy; they marry the cosmic with the mundane. Some are celestial objects millions of miles away, but you can hold parts of others in your hand, as meteorites. Scientifically, they are fascinating remnants of the earliest days of our solar system, and learning about them plays a small part in answering the big question of where we came from. But because of the hazard asteroids can pose, understanding these objects has a utilitarian purpose.

And, as Bob McMillan told me, searching for asteroids has other perks. "The best moments are when I step out of the observatory in the early-morning twilight at the end of an observing run and see the beautiful gradation of colors of the predawn sky along with sometimes bright stars or planets. Having finished the hard work of observing for several nights, that is a fine reward."

Seattle Public Library
Broadview Branch
(206) 684-7519

07/07/19 01:14PM

Borrower # 1350204

steroid hunters /
010090037911   Date Due: 07/28/19
cbk

TOTAL ITEMS: 1

Visit us on the Web at www.spl.org

6 My Favorite Telescope

Most asteroid hunters use telescopes sensitive to the wavelengths of light that our eyes can see. But there's one asteroid-hunting telescope that uses a different type of light. That telescope is called NEOWISE, and it is my favorite. I'm part of the team of scientists that gets to use it to find and study asteroids.

NEOWISE sees the sky in the thermal infrared, which means it sees heat. To understand how that works, you need to know that all warm things glow. You've probably seen the red glow of molten glass or metal, which are so hot they glow at visible wavelengths. As it turns out, things at cooler temperatures also glow, but at infrared wavelengths human eyes can't see. If you could see your surroundings right now in the infrared, you'd see that everything around you is glowing.

So when NEOWISE spots an asteroid, it's detecting the heat that asteroid is emitting. This gives NEOWISE a key advantage. Some asteroid surfaces are as dark as coal and reflect very little light. This makes them very hard to see if you're looking in visible wavelengths. But all asteroids are warmed by the Sun, making them glow in the infrared and shine brightly for NEOWISE.

The other nifty thing about looking at asteroids in the infrared is that you can measure how big they are. If you look at an asteroid in visible light, you cannot tell if it is a small, bright object or if it's big and dark. They both could reflect the same amount of light.

However, the amount of thermal light that an asteroid emits scales with size; the bigger it is, the more thermal light it will emit. So with NEOWISE, we can actually measure the sizes of asteroids. Although it seems like one of the first things you might learn about an asteroid, we don't yet know the sizes of most of them. This is because most asteroids have not yet been detected in the infrared. Thanks to NEOWISE and other infrared telescopes, we have good size estimates for about one in every five known asteroids.

The thermal infrared is pretty great for studying asteroids. So why is NEOWISE the only full-time infrared telescope hunting asteroids?

Well, it's not easy to build an infrared telescope, and this is partly because everything on Earth is warm enough to glow in the infrared, including telescopes. The faint glow of a distant warm asteroid is hard to see if the telescope you're looking through is also glowing brightly. To prevent this, astronomers have to keep their infrared telescopes very cold, dimming this glow.

Additionally, there's the problem of Earth's atmosphere. Although great for breathing, the atmosphere can be very annoying to astronomers, as it blocks some infrared wavelengths from reaching the ground.

Astronomers can sidestep these problems by putting their telescope on top of a gigantic rocket and launching it off the planet. So that's where NEOWISE is, right now. Up in space, far above Earth.

There are lots of advantages to having an off-planet telescope. Space is cold, and that helps keep your telescope from glowing. Space is above the atmosphere, so NEOWISE doesn't have any annoying life-giving molecules blocking its view of the cosmos. And the skies are always clear—no pesky rainclouds to get in the way. And best of all, no daylight! NEOWISE's orbit was designed so that the telescope is always looking away from the Sun (Rule #1), so it is always nighttime.

It takes a tremendous effort from a large group of people to successfully build and run a space telescope. Although today NEOWISE has just a handful of full-time scientists, hundreds if not thousands of people were involved in its conception, design, and construction, and dozens still work on daily operations. Every element of the project, like the design of the sensors, or the processing of the data, requires teams with highly specialized expertise, and managers to make sure all the teams are communicating effectively. Telling the stories behind a single spacecraft could fill an entire book by itself. Because the data from a space telescope is so hard-won, it is handled carefully; it is processed and archived so that it can be easily accessed and used by the worldwide scientific community.

As of the time I'm writing this, NEOWISE is a full-time asteroid-hunting robotic space telescope. But, as it happens, NEOWISE was not designed to hunt asteroids. It was made to study stars as cold as the human body, and to look at the most luminous galaxies. It was designed to operate for six months, and it did its job perfectly, under the leadership of astronomer Ned Wright, a professor at UCLA.[10]

But the astronomers saw more potential in this little space robot and wrote a proposal to NASA to use it for asteroid

Asteroids range in color and size, but even the darkest radiate heat NEOWISE can see.

NEOWISE is an asteroid-hunting telescope that scans the sky high above the Earth. This special vantage point means NEOWISE can look for asteroids 24 hours a day, and it's never bothered by clouds, the moon, or sunlight.

hunting. That proposal was approved, and the mission was extended. Six years later, NEOWISE is still ticking along, like a trusty used car. The cryogen that used to refrigerate its sensors is long gone, so we joke that the air-conditioning is broken. Operating for years beyond its designed lifetime, NEOWISE has racked up over a billion kilometers on its odometer as it orbits Earth. It's taken millions of images and has spotted hundreds of thousands of asteroids.

Asteroid hunting can be thrilling. But in many ways it is like what I imagine dairy farming is like: a job with constant responsibilities. The cows have to be milked and fed every day. The cows don't care if the weather is bad, if it is a holiday, or if you are sick. The cows need to be cared for.

Like cows, asteroids don't care if the weather is bad, if it's a holiday, or if you are sick. NEOWISE is a robot, so it doesn't care, either. It snaps a photograph of the sky every eleven seconds. And every day, NEOWISE takes some time to beam its data back to Earth. The data is beamed from NEOWISE to a network of relay satellites, to telescopes in White Sands, New Mexico. White Sands then sends the data to the Infrared Processing and Analysis Center at Caltech, in Pasadena, California, where I work.

Three times a week the data is analyzed by a supercomputer to find asteroids and comets. Day to day every star in the sky stays in the same location relative to each other. But day to day, asteroids will appear to move between the stationary stars. Our software is very good at finding asteroids and comets. If a moving object is found precisely where a previously discovered asteroid or comet was predicted to be, it collects those observations for us to submit to the Minor Planet Center.

However, if it finds a moving object where there aren't any known objects, it collects the images for human reviewers. Despite the immense advances in technology, human eyes and brains remain supreme for determining if a candidate set of images shows a real asteroid or some accumulation of artifacts, like cosmic rays. So three times a week everyone on the team pitches in to take a look at what the supercomputer has found.

Usually this is a fun and even relaxing task, something you can do while listening to music. But it is time-sensitive. If we find a new asteroid, we want to report it to the Minor Planet Center promptly, so other astronomers can look for it that night. Unlike dairy cows, asteroids don't have bells around their necks, so we don't want any of them to wander off before we get a good look at where they are going.

The images we are looking at are probably very different from what you imagine. When most people think of images from space telescopes, they think of the grand vistas returned by Hubble. And don't get me wrong—when combined and processed, NEOWISE images can be just as stunning.[11]

But stunning images don't help us hunt asteroids. Asteroids are very small, astronomically speaking, and very far away. So, as it was in Piazzi's day, asteroids look like stars in our images. That means they all, even Ceres, appear as points of light.

We want to find every single asteroid we possibly can, in each and every image. So the computer, when it is searching for asteroids, identifies the faintest points of light it can find. When we look at an asteroid the computer has found, we actually are looking at a tiny portion of a larger image. The computer has zoomed in and cropped around the spot where it thinks there is an asteroid, producing a pixelated grayscale square with a center patch of bright pixels that are only marginally brighter than the pixels surrounding them.

And no matter how fancy the telescope, this is what the newly discovered asteroids will look like—because we'll always want to find the faintest asteroids that telescope is capable of seeing. Like all astronomers, we want to squeeze every last drop of data out of our images.

So hunting for asteroids is a lot less like gazing at a beautiful color photograph and a lot more like reading a sonogram. Have you ever had a pregnant friend show you a very early sonogram of her baby, when the fetus is just a few pixels? And you've looked and looked, but the whole thing looks like a bunch of black-and-white noise? She's pointing at something, but you can't tell which smudge she's excited about. It's getting awkward, so you nod politely and congratulate her.

I've been with friends when I've seen the images of an asteroid that, until that very minute, was unknown to humanity. "There it is!" I've said. "We've found a new one!"

"Where?" they've asked.

"Right there! See that little bright smudge?"

They nod politely. "Very nice. Congratulations!"

I remember one asteroid discovery clearly. In spring 2014 I was sitting in a shared office at Caltech with my colleague Joe

Masiero. The office looked like many spaces used by scientists—slightly disorganized, with posters from old conferences on the walls. Two whiteboards were covered in numbers and equations, with a big note: "DO NOT ERASE." Streamers draped from ceilings and walls, left over from a birthday celebrated several months earlier.

The two of us were looking at the supercomputer output, and we knew the other NEOWISE scientists were doing the same in their offices. Joe said, "That's interesting!" and called me over.

It was a new asteroid, but we could both tell this one was a little special. For one, it was much brighter than most asteroids we discover; some are so faint that we cautiously wait until another observer has been able to spot it before we're convinced a new discovery was made. Not so with this asteroid: it was bright and clearly real. The other thing was that its location in the sky was very far south. Most asteroids tend to orbit the Sun in the same flat plane that the Earth does; this one had a tilt to its orbit (the technical term is *inclination*).

This new asteroid currently has the mellifluous provisional designation 2014 HQ$_{124}$; this type of designation is assigned to newly discovered objects. (After it has been observed long enough, it will get a number, and we'll get to name it.) The next step was to get more observations so its path through space could be calculated accurately. Unlike the Catalina Sky Survey, where observers choose each night where to point their telescopes, NEOWISE has a fixed scanning pattern. We needed telescopes on the ground to get these next observations.

We reported the observations to the Minor Planet Center, and astronomers in Southern Australia and New Zealand spotted it and also reported their observations to the Minor Planet Center. As observations of this object accumulated, it was soon clear that HQ_{124} was going to make a close approach to Earth. In fact, it was going to get within 1.2 million kilometers (750,000 miles). That's about three times the distance between the Earth and Moon.

Now, close approaches like this aren't particularly unusual; and the more asteroids we discover, the more of these approaches we are aware of. But each is a great opportunity to see an asteroid up close, and the scientific community mobilized to take advantage of it. With the NEOWISE data, we were able to measure its size, about 300 meters (1,000 feet). For context, if you somehow placed 2014 HQ_{124} in the middle of Paris, it'd be about as tall as the Eiffel Tower (and a lot wider). Other astronomers used special instruments to measure the amount of light it reflected at different wavelengths. This technique, called spectroscopy, lets astronomers identify the minerals on the asteroid's surface.

This was also a great opportunity for a unique technique to shine: interplanetary radar. Buckle up, friends, this is pretty cool. You're probably familiar with military radar, where technicians send pulses of radio waves out, and listen for any to bounce back. If some bounce back, you know they are bouncing off something, like maybe a missile. And by the way they bounce back, you can tell where the missile is and how fast it's going.

Turns out you can do the same thing with asteroids. But, as you can imagine, you can't do this with just any radar setup; you need an extremely high-powered radio transmitter, and a particularly big receiving dish.

Fortunately, there's a few observatories in the world that can do this type of work. One is in Arecibo, Puerto Rico. This is a big white dish, a radio telescope, set in a large natural depression in the middle of the jungle. One of the world's largest telescopes, it is 305 meters (1,000 feet) across. Another is a set of large radio telescopes called Goldstone in the California desert. They aren't nearly as big as Arecibo, but are still pretty large; the advantage to these telescopes is that they can swivel and see a broader swath of sky.

Because it relies on detecting an echo from an object, the ability of radar to detect an asteroid depends on both how big the object is and how far away it is from Earth. The radar signal spreads out as it travels, getting weaker the farther it gets from Earth. A larger object can reflect more of the beam, and therefore be detected when it is farther away. For example, radar has been used to study Jupiter's moon Europa. Although it depends on the relative positions of Earth and Jupiter, it always takes more than an hour for the radio waves to travel to Europa and bounce back.

A small asteroid isn't big enough to create an echo we can detect here on Earth unless it is pretty close. (This is for astronomical definitions of "pretty close"—millions of kilometers/ miles.) And the closer the asteroid, the stronger the echo, and the more information we can get out of it. Radar astronomer Lance Benner likes to put the capabilities of this technique into perspective by saying that radar could detect a golf ball out at the distance of the Moon.

In order to observe an asteroid using radar, you need to know exactly where it is going to be. Unlike NEOWISE or Catalina's telescopes, which can see large swaths of sky at a time, the radio-wave beam, by necessity, is narrowly focused. Since NEOWISE discovered 2014 HQ_{124} several months before its close approach, there was time to get enough observations to predict where the asteroid was going to be precisely enough for the radar astronomers to catch it with their narrow beams.

But HQ_{124} presented a particular challenge. It was going to be so close to Earth that the time it would take for a radio wave to be transmitted, bounce off the asteroid, and return was only nine seconds. A telescope could switch between transmitting and receiving mode that quickly, but astronomers knew they could get better data if they used two antennae to observe it, one dedicated to sending and another to receiving radio waves.

Luckily, the radar astronomers had a clever technique at their disposal, one that used two radio telescopes. Jet Propulsion Laboratory astronomers Marina Brozovic and Lance Benner coordinated the transmission of radio waves from Goldstone, while Arecibo astronomers Michael Nolan and Patrick Taylor used their telescope to "listen in" for the asteroid-bounced echo.

This was the first test of new data-taking equipment at Arecibo, and it worked perfectly. The images have 3.75-meter resolution and revealed depressions and a boulder on a bowling pin–shaped asteroid surface.

2014 HQ_{124} holds a special place in my heart, as I was there to witness it from discovery to close approach, and I'm sure I'll see it again in the sky. But although it's special to me, it isn't a particularly unique asteroid. There are thousands of asteroids that big, and it isn't uncommon that one gets that close to Earth. And after all, there are many more to be discovered.

NEOWISE is unique among asteroid surveys; it sees in the infrared, and its vantage point from Earth orbit allows it to see the entire sky. These qualities enabled astronomers, using NEOWISE data, to answer a very important question: How many asteroids are out there?

It's a puzzling question. How do we possibly know how many asteroids we haven't found, if we haven't found all of them yet? How do we know what we don't know? For the answer, we turn to a clever technique that astronomers have used to answer all sorts of questions: debiasing.

As you can guess from the name, this technique hinges on eliminating biases. Visible-light telescopes are biased toward detecting asteroids that have bright surfaces that reflect a lot of light. Ground-based telescopes are biased in that they can only see the part of sky that is above the telescope, and only during clear weather. And many places have more clouds during one part of the year; winter storms in the North, or summer monsoons in Arizona, for example. With seasonal storms, you can miss parts of the solar system on a fairly regular basis. Every month when the Moon is full, its extra light prevents the discovery of the dimmest asteroids.

However, NEOWISE was in a unique position to not have these biases. Its infrared sensors see heat, so light and dark asteroids show up equally well. Its orbit around the Earth allows it to see the entire sky every six months, and it pivots to avoid any moonlight. There are no clouds in space. It's an ideal experiment.

To determine how many asteroids were left to be discovered, the NEOWISE team, in a study led by Amy Mainzer, constructed a computer simulation of the first portion of NEOWISE's survey. They knew exactly how sensitive NEOWISE's detectors were, and knew the smallest objects NEOWISE could find. They also knew the time and location in the sky of every image NEOWISE had ever taken. The computer simulation exactly modeled how NEOWISE observed the sky, and what it would be able to see.

Then, they simulated hundreds of thousands of "synthetic" asteroids and ran them through the simulation to see how many asteroids NEOWISE would have seen. This result was then compared to what NEOWISE actually saw.

Some people find this concept a bit counterintuitive, so let's look at a simple example. Let's say Amy generates fifteen asteroids of a certain size and type of orbit. She runs those asteroids through the NEOWISE simulation and learns that if there were fifteen asteroids out there, NEOWISE would have seen ten. She then looks at the actual number of asteroids of that size and type of orbit that NEOWISE saw. Let's say that NEOWISE actually saw twelve. She goes back and adjusts her simulation. Maybe this time she generates eighteen asteroids of this type and runs the simulation again. This time, the simulation says NEOWISE detected twelve asteroids. Bingo! Amy knows that the total number of asteroids out there is eighteen, and there are about six asteroids of that size and type still waiting to be found.

Of course, the actual implementation is more complicated than that, and many more asteroids are simulated so that the results have statistical significance. But you get the idea. With this method, we now know what we don't know. Many people are particularly interested in the class of asteroids that get particularly close to Earth. As of the publication of the study in 2012, over 90 percent of asteroids bigger than one kilometer across had been found, but only about 30 percent of all asteroids bigger than 100 meters across in this class had been discovered. We now know what we have left to find.

7 The Giggle Factor

In 1994 a collection of academic papers was published in a book called *Hazards Due to Comets and Asteroids*, edited by Tom Gehrels. As you might imagine from the title, these were papers written by scientists and policymakers interested in the consequences of an asteroid or comet hitting Earth. Today, over twenty years later, it's still a fascinating read. While many of the scientific questions and issues remain relevant, it's inevitably a product of its time. Reading it is like watching a '90s sitcom—you relate to the characters and laugh at the jokes, but you suddenly notice how much time has passed when someone pulls out a laptop the size of a suitcase.[12]

"The Comet and Asteroid Impact Hazard in Perspective" is a chapter written by Paul Weissman, then a scientist at the Jet Propulsion Laboratory. This is how it begins: "The potential hazard from comet and asteroid impacts is one of a number of serious natural and man-made calamities facing modern society. However, only three of these currently have the potential to wipe out a significant fraction of human life on this planet: impacts, nuclear war, and the AIDS epidemic."

When I read that line, it stopped me in my tracks. Today, AIDS is a devastating disease, responsible for more than a million deaths a year. But in the early '90s, AIDS seemed like something else entirely—an uncontrollable, perhaps unstoppable, pandemic. In the chapter, Paul explains,

"The AIDS epidemic has now spread worldwide; an estimated 10 million people are infected with the AIDS virus. . . . Intensive medical research efforts to develop a cure and/or vaccine have so far produced only limited results. It is entirely possible that a solution may appear at any time, but at present the disease continues to spread at an alarming rate."

Today, our perspective is quite different. AIDS is no longer on par with nuclear war. What was once a seemingly insurmountable problem now appears more and more fixable; in part thanks to the hard work of medical researchers and policymakers.

I mention this because I think it is important to remember that even big, terrifying problems can be solved. We have a ways to go before AIDS is eradicated, but it no longer seems to have the same potential for destruction as it did in the 1990s. Likewise, the threat of an asteroid impact is not some existential threat beyond our control; as you will see, it is actually a solvable problem.

But in 1994, Paul, like many scientists of the time, was concerned that the threat of a comet or asteroid impact wasn't taken seriously. This was often referred to as "the giggle factor."

Don Yeomans, another noted comet scientist, recalled in an interview for a 2013 *Nautilus* magazine article, "People would laugh and say, 'Yeah, when was the last time?' Simply because we didn't see them, they didn't take the threat as seriously as we have come to."

Most government officials were also unconcerned. Another chapter[13] in *Hazards Due to Comets and Asteroids* is written by Robert Park, Lori Garver, and Terry Dawson, who were all involved with policy-making. They explained: "Clearly, elected officials in Washington are not being inundated with mail from constituents complaining that a member of their family has just been killed or their property destroyed by a marauding asteroid. Indeed, the prevailing view among government officials who hear about this issue for the first time is that the epoch of large asteroid strikes on Earth ended millions or billions of years ago."

But public perception was about to rapidly change. Between when Paul wrote his article and when it went to press, some-thing big happened: Comet Shoemaker–Levy 9 slammed into Jupiter.

Shoemaker–Levy 9 was discovered the year before, in 1993, by Carolyn and Eugene Shoemaker and David Levy. It looked unusual from the start; it appeared to be fragmented into sev-eral pieces. Further observations led to an orbit, which revealed that the comet had been caught by Jupiter's gravity and ripped apart in 1992. Now trapped, it was metaphorically circling the drain and was predicted to impact Jupiter in 1994.

Jupiter is a truly enormous planet; if it were hollow, hun-dreds of Earth-sized planets could fit inside.[14] And on top of that, it is a ball of gas; the exterior is miles and miles of cloud layers. Many scientists were skeptical that these relatively tiny comet fragments (the biggest was 4 kilometers, or 2.5 miles, across) would do much of anything. Perhaps the comet was just a loosely packed "snowball" of ice and dust, and Jupiter's gravity would disintegrate it before it entered the atmosphere. Or perhaps it would just plop into the planet like a bowling ball dropped through some fog, a brief event that would be quickly smoothed over.

However, it was clearly a rare opportunity, and everyone wanted to see what would happen. Unfortunately, the impact was going to hit the side of Jupiter that was facing away from Earth, so astronomers with telescopes wouldn't have a direct view. A few clever amateur astronomers decided to watch Jupiter's moons during the time of the impacts, using their icy surfaces as mirrors, hoping they'd reflect any flash of light that might occur during impact.

A fleet of spacecraft was trained on Jupiter, including the Hubble Space Telescope, the *ROSAT* X-ray satellite, and the *Galileo* spacecraft, which just happened to be on its way to orbit Jupiter and could see the impacts as they happened. Even *Voyager 2*, then over 6 billion kilometers (4 billion miles) past Jupiter on its way out of our solar system, was programmed to listen for radio waves that might come from the impact.

The results were extreme. The first fragment of the comet plunged into the atmosphere at a speed of more than 200,000 kph (120,000 mph), creating a fireball seen by the *Galileo* space-craft with a peak temperature of 24,000°C (43,000°F). That's so hot it's hard to put into context with things we deal with here on Earth. Even an oxyacetylene welding torch is only 3,300°C (6,000°F). As the subsequent fragments impacted, they rang the planet like a bell, creating enormous waves in the atmo-sphere that traveled across the planet at 1,600 kph (1,000 mph).

They left behind enormous dark spots that persisted for months. The largest dark spot was so large it could fit the Earth inside it, and could be seen with a backyard telescope.

I was just in grade school, but I remember seeing the Hubble images of Jupiter, scarred by those dark marks, on the front page of the *Los Angeles Times*. The implication was clear: What if this had hit Earth?

People were starting to take the threat of impacts seriously, and Shoemaker–Levy 9 had a lot to do with that. But the Shoemaker–Levy 9 impact isn't exactly a household name. What are household names are the movies that came out in 1998, four years later—*Armageddon* and *Deep Impact*. Both were summer blockbusters involving an asteroid or comet hitting Earth.

It's hard to take these films too seriously; they are designed to entertain, not inform. And it's difficult to quantify how they have influenced public perception of the impact hazard. But I have a couple of suspicions. The first is the timing; both came out four years after Shoemaker–Levy 9. I wonder if the screenwriters didn't also pick up their copies of the *Los Angeles Times* and wonder what would happen if the comet had hit Earth instead. And my second suspicion is that, despite being fictional and even a bit ridiculous at times (looking at you, *Armageddon*), people walked out of the theaters with a fresh understanding that an impact with Earth could be pretty bad, and it didn't seem entirely unlikely.

Public perception changed more rapidly in the late '90s than anyone had predicted. And as the probability of an asteroid or comet impact started to be taken more seriously, more people began to consider what might be done to minimize the damage from such an event. In other words, how do we prepare?

The first thing to consider is that we can predict the orbits of these objects with precision and accuracy. Asteroid trajectories are governed by gravity and, to a much lesser extent, sunlight—two mechanisms we understand very well.

I asked Jon Giorgini about current orbit prediction capabilities for near-Earth asteroids. Jon works at the Jet Propulsion Laboratory curating a system called Horizons that keeps track of every known object in the solar system. He said, "If it's newly discovered and we've just seen it for a few months, on average we can predict those ahead about eighty years—if you have only optical measurements. If we can get radar measurements, that goes up by a factor of five, so you can predict ahead, on average, about four hundred years. That's just for an object that's been recently discovered. If we've been tracking it for years and it has gone around its orbit more than once, we can predict [orbits], on average, about eight hundred years into the future."

Although Jon states these facts with a casual air, they always impress me. Using the language of mathematics, we can figure out exactly where these asteroids will be for each and every day between today and several hundred years in the future. I can't think of much else we can predict with such certainty. In 2400, the world will likely be unrecognizable to us now; new countries, or perhaps no countries at all, language and culture will have changed, but Jon's orbital predictions of tens of thousands of asteroids will still be spot-on.

We have the observational tools and computational ability to predict exactly where these objects are going. It's a vastly simpler problem, mathematically, than predicting the path of a hurricane weeks in advance, or when an earthquake will happen.

The second thing to consider is that, given enough warning, we have the technological capability to deflect an impactor. Although we haven't ventured far from our home, we are a space-faring species. Park, Garver, and Dawson reported that a 1992 workshop of experts concluded that "Available technology can deal effectively with threatening asteroids, given warning time on the order of several years." Available technology has advanced considerably since then; however, the warning time of several years, or even a decade, is still a necessity.

These deflectors could take many forms, and the right tool for the job would depend on the specifics of the situation. One option is to hit the asteroid with something heavy, a good shove that would put it on a slightly different course. Another is to send a spacecraft to orbit the asteroid and then use the gravitational pull between the asteroid and the spacecraft to gently tug the asteroid off a collision course. And under certain circumstances, nuclear detonations would also be considered; practically, they remain one of the most effective ways to release a tremendous amount of energy (more on this later).

Since hazardous impacts are very rare, it doesn't make much sense to have a standing arsenal of deflection mechanisms built in advance. Not to mention this would be hugely expensive. Park, Garver, and Dawson wrote, "It is unrealistic to expect governments to sustain a commitment to protection against a rare occurrence when they are constantly under pressure to respond to some perceived immediate crisis."

We have the technology to deflect an asteroid, but to do that we'd need time. Since it isn't practical to build an asteroid deflector in advance, we'll need to make sure we have an early-warning system that can tell us if one is heading our way.

The best thing we can do now to deal with this hazard is to search the skies and discover as many asteroids as we can. Hopefully, an extensive search may find there's nothing headed our way for hundreds of years, which would be quite nice. But we need to know if that's not the case.

So the good news is, we do know how to find asteroids, and we're very good at it. These three factors—that asteroid paths are predictable, that we know how to find them, and with enough time we can prevent a collision—combine to mean that an asteroid collision with Earth, despite appearing to be one of the most uncontrollable and terrifying of all possible disasters is, paradoxically, actually one of the most predictable and preventable.

As with all natural disasters, the more we learn about them, the more it is clear that preparation can keep us safe. This is not something to be feared, only something to be prepared for. And preparation, in this case, means watching the sky.

Since early detection is key, we should all take a moment to thank some little-noticed US legislation, known as Spaceguard, which directed NASA to start the Near-Earth Object Observations Program in 1998. Motivated by a report headed by scientist David Morrison, this was a congressional directive to discover at least 90 percent of one kilometer or larger asteroids with orbits near Earth's within ten years. It led to increased funding of asteroid surveys, which began finding huge numbers of asteroids using CCDs. In 2005, the George E. Brown, Jr. Near-Earth Object Survey Act expanded the goal to find 90 percent of objects nearby Earth 140 meters or bigger by 2020. That's the target asteroid hunters are working toward right now.

When the major surveys began being funded, the number of known asteroids grew exponentially. As we hunt and find more and more asteroids, we are more aware of when they make close approaches to Earth. These flybys often make the news, and today asteroids are something most people are aware of. The "giggle factor" is gone. As Don Yeomans put it, "Nobody's laughing now."

8 The Planetary Defense Coordination Office

On the whole, asteroids tend to leave Earth alone. And day to day, there really isn't cause for most people to give them any thought. But despite being remote objects that most people have never seen, asteroids, and the threat they pose, come up an awful lot in popular culture. They are mentioned offhandedly. "Maybe I don't have to do my homework—an asteroid could hit the school tomorrow." They are plot drivers in movies and TV shows. I recently asked Lindley Johnson what his favorite depiction of an asteroid impact was in popular culture.

"The movie *Meteor*. It came out in the late seventies, and it starred Sean Connery, who played me."

Yes, Lindley Johnson has a job that not only sparks the imagination of Hollywood scriptwriters but also inspires studios to cast a 1970s Sean Connery for the part. Lindley is NASA's Planetary Defense Officer. Working at NASA's headquarters in an unremarkable office building in Washington, DC, he manages almost everything the US government does related to near-Earth objects. ("Objects" here refers to asteroids and comets.) He directs funding to the surveys, the computational centers that calculate orbits, and scientific research into these objects. If asteroid hunting in the USA were a business, he'd be the CEO.

As Planetary Defense Officer, he heads NASA's Planetary Defense Coordination Office. He explained, "The word *coordination* is important. We're not doing everything. Our office helps to coordinate all the effort by US government agencies and a lot of international work that's going on toward detecting the threat, and figuring out what we would do if there was an asteroid on an impact trajectory." He leads a team of experts who work with the UN Committee on the Peaceful Uses of Outer Space, and plays a major role supporting the International Asteroid Warning Network, which coordinates observatories across the world.

Obviously, this position comes with tremendous responsibility. Although it's fun watching Sean Connery play the action hero on-screen, I think few of us would actually want the pressure that comes along with the real-life job. I asked Lindley about this, and he responded in his characteristic calm cadence (he sounds the same whether he's testifying before Congress or ordering a coffee): "I think it is a big responsibility; however, not one that I find particularly high-stress. I don't know but part of it is probably my air force training and experience. I've been in a lot of higher-stress jobs where things required more immediate reactions."

And that's the key to the whole issue of asteroids hitting Earth—time. If we can find something with enough early warning, we have the technological prowess to deflect it. By searching the sky now, we hope to give ourselves the early warning (ideally decades) that we need.

But that strategy of focused searching hasn't stopped people from thinking about what we might do if an asteroid was on its way toward us.

The consensus of the community, Lindley told me, is that current technology provides three viable deflection techniques.

"One is a tried-and-true method; hit it hard to move it in a different direction, the so-called kinetic-impactor technique," he said. "A sufficiently sized impactor, of sufficient velocity in the right direction, hits the asteroid hard enough to change its velocity by a few millimeters to centimeters per second, far enough ahead of the predicted impact. This changes the time when it would arrive, making it a miss instead of a hit."

This is a simple technique, and that simplicity is appealing. To deflect an asteroid, you want to use a method where you understand as many of the variables as possible, so that you can predict the results accurately. Changing the velocity changes the time when the asteroid crosses Earth's orbit. After all, just because an asteroid crosses Earth's path doesn't mean there is necessarily going to be a collision. The asteroid has to cross Earth's path when the Earth is right there. And the Earth is moving pretty fast, about 30 kilometers per second (18 miles per second). So change the asteroid's speed a little bit, and it zooms by a little early or a little late. Problem solved.

NASA has hit a comet with an impactor, during the Deep Impact mission. The goal of that mission was to study the surface by making a crater and stirring up the surface material so it could be studied. To do that, the spacecraft hit the comet with something heavy: a 372 kg (820 lbs) probe traveling at 37,000 kph (23,000 mph). That's faster than a bullet.

The mission's goal wasn't to change the comet's orbit, but as you might expect, the orbit did change, very slightly. The impact

NUCLEAR OPTION
If time is short, a nuclear detonation could knock the asteroid off course.

GRAVITY TRACTOR
Use the gravitational tugs of a spacecraft to slowly bend the asteroid's path.

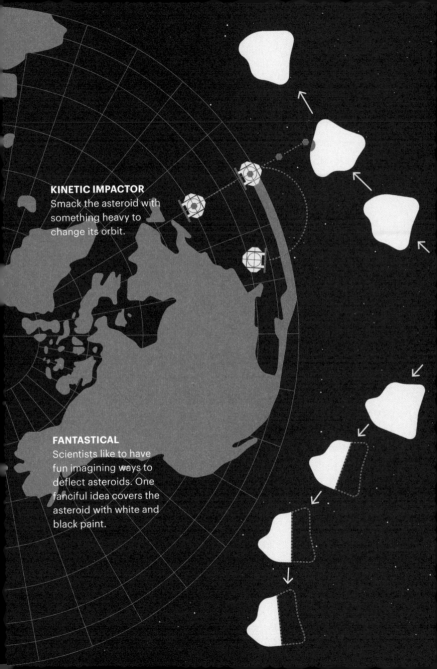

KINETIC IMPACTOR
Smack the asteroid with something heavy to change its orbit.

FANTASTICAL
Scientists like to have fun imagining ways to deflect asteroids. One fanciful idea covers the asteroid with white and black paint.

altered the comet's velocity by about 0.00005 mm/s. To put that in perspective, at a speed of 0.00005 mm/s, it would take you seventy-two hours to traverse the surface of an M&M candy.[15]

To understand why such a massive probe, traveling faster than a bullet, caused such a small change in the comet's speed, I like to think of things in terms of really big pumpkins. The impacting probe, weighing in at 372 kg (820 lbs), is about as big as one really big pumpkin (though, fun fact, record-setting pumpkins can top 900 kg, or 2,000 lbs). The comet is about as big as a ball of 300 billion really big pumpkins. Even though the giant-pumpkin impactor was traveling faster than a bullet, when it smashed into the 300-billion-pumpkin comet, the comet remained mostly intact.

So if we wanted to deflect an asteroid, we could launch a spacecraft that would hit the asteroid with something heavy. But that something heavy, even if it was traveling very fast, would have to be a pretty significant fraction of the asteroid's mass. We would need something quite a bit heavier than a very large pumpkin.

There are other complications to the kinetic-impactor deflection technique. Some asteroids seem to be mostly solid chunks of rock or metal. But some appear to be very loosely held-together clumps of dust, boulders, and small rocks. Gravity holds them together, but gravity on these small objects is almost impossibly weak; roughly 1/100,000 of what it is on Earth (assuming an object about a kilometer across). Get one of these asteroids—descriptively called "rubble piles"—and a kinetic impactor might not deliver much of a punch; its energy might be absorbed by the interior, like a bowling ball being dropped on a beanbag. Therefore, any deflection technique would have to be tailored to the individual asteroid. So one good thing to do now is study asteroids so we can know what types of objects we might have to deal with.

But, there are other options, as Lindley explains: "Another technique that is talked about is the gravity tractor, where the minute mutual gravitational attraction between an asteroid and a spacecraft has a very small effect."

In this case, a spacecraft is sent to the asteroid and begins asymmetrically orbiting around it.

"The spacecraft is positioned to maneuver in a determined direction, to slowly tug the asteroid off its natural trajectory and onto a trajectory where it would be a miss instead of a hit."

This is an elegant technique; it is gradual and exquisitely controllable. As the small gravitational tug of the spacecraft reshapes the asteroid's orbit, engineers and scientists could carefully monitor the change and make adjustments. And, even better, you never need to actually touch the asteroid. A fast-spinning asteroid could be moved in this manner. Even a rubble-pile asteroid could be towed while remaining completely intact.

The disadvantage, though, is that the gravitational pull of a spacecraft is very, very small. Even if we built a spacecraft bigger than any we have ever made, this method would take time to implement, especially if the asteroid itself was big. We could try to increase the mass of the spacecraft after it gets to the asteroid by having it grab large, heavy boulders off the asteroid's surface, in a scheme called the "enhanced gravity tractor." But that would be technically challenging, and may require more time to design and build. Either way, we'd need a lot of warning, ideally decades.

But there's also a third seriously considered technique.

"We call this the nuclear option, and it's one we really do not want to use. But if time is short, and particularly if the object is a larger size, the only source of energy mankind knows of that is sufficient to shove an asteroid off an impact trajectory would be detonation of a nuclear device.

"It wouldn't be designed to blow it up into little pieces, as Hollywood movies show, but would instead irradiate an area of the surface, heating it up instantly and causing it to blow off, thereby imparting a force at the right time and in the right direction, again turning a hit into a miss."

In other words, the bomb is detonated next to, not on, the asteroid. The explosion is designed to give the asteroid a hard shove, not blow it up into a bunch of pieces.

A nuclear explosion is powerful and makes for a good movie, but there's a lot we don't know that makes it hard to predict exactly what would happen after such a detonation. How would the asteroid surface react? If the asteroid was spinning quickly, it might be hard to time the detonation so that the explosion delivered a good shove, and the energy didn't just go into speeding up or slowing down the asteroid's spin. A grapefruit-shaped asteroid would react differently from an eggplant-shaped one. And how would a rubble-pile asteroid react? Would it separate into several pieces, only to reaccumulate back together? In a serious situation, you want to be able to predict the outcome precisely, and there's more unknowns with this technique than we would like.

These are the three methods currently being seriously considered. But the idea of asteroid deflection has inspired several more creative approaches. One of these inventive ideas is that you could paint one side of an asteroid black and the other white. This would mean the two sides of the object would absorb sunlight differently—the white side would reflect the light, and the black would absorb it, warm up, and release photons a little while later in the form of infrared light. Those departing photons would carry away momentum. In practice, it would be like having a tiny, tiny rocket thruster. But this would likely be far too weak a force, and far too slow, to be very practical.

People have also thought of using chemical rocket thrusters to move the object, but they wouldn't work well if the asteroid was spinning quickly. There's also been a proposal to use magnetic coils to electrostatically grab loose dust and small rocks from the surface, like hair sticking to a charged balloon. Once the spacecraft collected the dust, it would throw it back at the asteroid at high speed, using the asteroid's surface material to give it a shove. There's a bit of poetic justice to this technique that I find appealing.

Even though none of the asteroids we know about are going to hit Earth, experts are seriously considering how we, as a species, might react. The best course of action depends on what the asteroid is made of (rubble pile or solid chunk of metal), its particular path around the Sun, the way it is spinning (some asteroids spin slower than Earth, while others make a full rotation in less than a minute), and how much time we have before a predicted impact. With all these unknowns, the best plan would have to be tailored to the individual situation.

It pays to be prepared. Being prepared means looking for new asteroids and learning about the ones we've already found. By trying to discover as many asteroids as possible now, we hope to give ourselves as much early warning as we can if there does happen to be something big headed toward us. Likewise, by studying as many of the known asteroids as we can, we are trying to get a sense of the types of objects we might have to deal with.

In the meantime, the international community of scientists and government officials also prepares by conducting exercises, like war games, but with a hypothetical asteroid. Experts from across the world participate, including representatives from the US Federal Emergency Management Agency (FEMA).

I asked Lindley to describe what might happen if an asteroid 150 meters (500 feet, roughly as big as a cruise ship) across was predicted as having a chance of hitting Earth ten years in the future.

"The first thing that's got to be done is to really understand as best we can what the orbit is, and if it's really an impactor or not," he said.

To determine its orbit, astronomers would need as many observations of the asteroid's position as possible. This would require international cooperation. Some asteroids are only visible from one hemisphere on Earth; it's possible that tracking would need to be done by astronomers in countries below the equator. And since telescopes can't look at the Sun, you'd want observatories at different longitudes, so the asteroid could be monitored constantly as night falls around the globe.

But even after thousands of observations, there's still going to be a small amount of uncertainty in where exactly the asteroid is going. Although the orbit might be predicted with extraordinary precision, the chance it has of hitting Earth is expressed as a probability. This is very similar to predicting the weather. Meteorologists can see a storm system headed toward your city, but a lot of the time they can't predict if your particular house will get rain. Just as a weather reporter might say you have a 20 percent chance of rain, astronomers will compute a probability of impact.

Lindley continued, "If the probability of impact were high enough—and there's some debate on what that means—one percent? ten percent?—we'd have to decide at what point we would need to start to take action to accomplish a deflection mission.

"It would be multiple missions, probably multiple nations involved. It becomes as much of an international coalition-building effort as a technology effort. I think responding to a potential asteroid impact would be as complex an endeavor as mankind has ever undertaken."

This is the type of scenario most people imagine, thanks to movies like *Armageddon* and *Deep Impact*. But depending on the circumstances, the response could be quite different. I asked Lindley what it might look like if a twenty-meter (roughly convenience store–sized) asteroid was predicted to hit Earth two years from now.

"Well, in that case, a twenty-meter object cannot do a lot of damage, although Chelyabinsk in February 2013 showed us it is not to be trifled with. Again, we would firmly establish an orbit as best we could, to determine the impact time and where the impact site would be as precisely as we could."

If the object was to hit over a populated area (which is statistically unlikely), the residents could be advised to "shelter in place," to stay inside, away from windows. The amount of damage a small object like this could do would depend on what it was made of and how it was expected to break up in the atmosphere.

"Researchers are hard at work trying to understand the effects the passage through the atmosphere has on objects of various compositions and sizes," Lindley said. "How big do they have to be to survive relatively intact through the atmosphere and cause impact craters? What would be the radius of the effects from such an event?"

For an object a bit bigger than twenty meters across, it might make the most sense to evacuate the area it was expected to impact.

"Below some size threshold and composition, it's not worth nations trying to undertake a deflection mission. It makes more sense to determine as best we can where the impact site will be and evacuate."

Evacuation, though not very flashy, might be the best way to protect people from harm. The most sensible option might be just to get out of the way.

When we imagine an asteroid impact, I think we all picture a dramatic montage as the leaders of countries are informed, one by one. I wondered if Lindley would ever call the president.

"Calling the president is a euphemism," he said. "There's not that many people who actually have the president's phone number."

I was a little disappointed. But what Lindley said next illustrates the part of his job that Sean Connery doesn't portray—the unglamorous business of working within the chain of command of a large bureaucracy.

"In our notification plan, when we detect an impactor, we do up-channel that information to the White House. The speed of that communication is directly related to the time until impact. If it's predicted to impact twenty-five years from now, we certainly aren't going to wake up the president to tell him about it tonight. It will take a few days to fully verify and validate the information before we make such a notification, if it is years away. But if it's only a few days or hours then that information needs to get up to the leadership level as quickly as we have confidence of what we are talking about . . . and indeed that may be a call from the NASA Administrator to the White House."[16]

This is where the science meets policy-making. After all, scientific consensus doesn't do much to change the world on its own; it takes people to create, fund, and manage new courses of action. Lindley credits past scientists for bringing this issue to the forefront.

"Nothing would have gotten done in this arena at all if not for very dedicated astronomers and scientists in the previous generation: Carolyn and Gene Shoemaker, Eleanor Helin, Tom Gehrels, and David Morrison bringing this issue to the attention of decision makers.

"And then the current generation of asteroid hunters who, and you are one of them, dedicate their careers, a good part of their entire lives, to detection, tracking, and characterization of objects that could one day harm us. . . . A one-hundred- to two-hundred-meter object hitting us in the wrong place would be a disaster of greater scale than we've ever seen in the historical record."

Lindley has held this position for over a decade. Over that time his office and responsibilities have expanded significantly. I asked him to reflect on his work at NASA.

"I look at it more as, without wanting to sound too self-important, a 'service to humankind' kind of job. This is my service to the world. A lot of people work on relief efforts and charities and curing medical issues. Those kinds of things are a more immediate priority and affect people's lives on a daily basis. I have different skills, and I'm working on something I consider as important but may not show direct benefits for years to come. It's very much a legacy project."

I think of it as an insurance policy. The work we are doing now—hunting asteroids, learning about them, and thinking of how to move them—is all designed so that we can be as prepared as possible in case a threatening object is found. And although great progress has been made, the work is by no means complete. We have the technology to find all the potentially hazardous asteroids. Now we need to finish the job.

Something tells me that you probably picked up this book because you're concerned about an asteroid hitting Earth. Exotic, dangerous, and remote, asteroids certainly hold a special place in our cultural psyche.

But I hope that now, along with a clearer understanding of what asteroids are and how we hunt them, you feel a sense of relief that an asteroid impact is not only improbable but also entirely within our means to control. Absolutely, we must keep searching until we have found all the potentially hazardous asteroids. But if we can do that quickly and effectively enough, we will have given ourselves enough time to mount a defense if one is headed our way. Put another way, an asteroid impact could go down in history as the first natural disaster that humanity prevented.

We could also learn, after finding all the potentially hazardous asteroids, that all of them are actually on peaceful trajectories, avoiding Earth for hundreds of years. That would be pretty great. Of course, then we'd need to find something else to pin our cultural existential angst to (though I'm sure we'd come up with something).

Beyond the hazard, discovering asteroids is, at its heart, basic exploration of our environment.

In 1968 we first ventured to the Moon. Over the next decade, twenty-four astronauts made the trip. And in the minds of the public that crossing became routine, like traveling a well-worn path. But space is not static. Asteroids orbit around our Sun, and at least once a month a small one passes closer to Earth than the Moon. With asteroids still left to discover, we haven't completed the map of our cosmic backyard; a region constantly changing as Earth, Moon, asteroids, and comets all travel their particular routes through space.

I am grateful to play a small part in this effort, alongside my colleagues on the NEOWISE team. I often think of the search for asteroids as a giant public works project, an accomplishment that will last for generations, a testament to the power of teamwork.

Completing the map of near-Earth space will be a milestone for a space-faring species. Past mapmakers wrote "Here Be Dragons" to denote unexplored and possibly dangerous territory. Right now our map of near-Earth space still contains unexplored territory; there are dangers lurking in the dark places. But we are charting these regions, replacing these unknowns with discovered asteroids, one by one. We may be banishing the monsters from near-Earth space, but that doesn't mean there are no mysteries left. We've only got the barest information on most of these objects, and each is an unexplored world unto itself. We've only started exploring. Right now, somewhere, asteroid hunters watch the sky.

ACKNOWLEDGMENTS

I owe so many people thanks. First, my editor, Michelle Quint, for her guidance and excellent feedback, as well as Grace Rubenstein and Ellyn Guttman at TED Books. Enormous thanks to the TED Fellows team, led by Tom Rielly; without the efforts of these folks, this book would not exist.

I would like to extend a far-reaching thanks to the community of planetary scientists and asteroid hunters. They have built such a wealth of knowledge that I could barely scratch the surface here.

Heartfelt thanks to everyone who gave interviews. The US taxpayers who fund asteroid hunting. All the folks at NASA Headquarters. Lindley Johnson for granting an interview and providing valuable feedback on drafts. Tim Spahr for acting as a scientific reviewer of the manuscript. The NEOWISE team for being stellar colleagues.

I am grateful to my family for their unwavering support. And I am grateful to Robert for his wit, insight, and patience as I worked on this project.

1   I've also maintained that the asteroids are the real heroes of that movie. After all, it's the asteroids that take out the bad guys; the *Millennium Falcon* is just getting out of the way.

2   Asteroid scientists love comparing the asteroid field in *The Empire Strikes Back* to our own main belt. José Luis Galache did a back-of-the-envelope calculation on the same topic that appeared in the *New York Times*.

3   Michael was talking to me on my podcast; to hear more, listen to Episode 16 (October 25, 2016) of *Spacepod* (http://www.listentospacepod.com).

4   There are lots of numbers in this book. When a number is given in two different units, I've made sure each value has the same number of significant figures. Significant figures, which you might have learned about in high school chemistry, are used to communicate how well a value is known. In this case, "100 tons" is exactly equal to 90,718.5 kg. But when you read "about 100 tons," you get the sense that this isn't a precise measurement. The actual value might be 110 tons, or 92 tons, or 103 tons. On the other hand, if you read "90,718.5 kg of dust and small rocks hit Earth each day," you get the sense that we know that measurement precisely, down to the half kilogram. That's misleading. So instead, I wrote that 90,000 kg of dust and small rocks hit the Earth each day. That gives you a better sense of how precise these numbers are.

5   In fact, if you want to look at the Sun with a hobby telescope, you need a special filter that blocks 99.97 percent of the light.

6   To hear Tim Spahr explain this in his own words, see Episode 34 (February 28, 2016) of the *Spacepod* podcast (http://www.listentospacepod.com).

7   For an excellent history of this search, see the book *Asteroids III*, "Giuseppe Piazzi and the Discovery of Ceres," by G. Foderà Serio, A. Manara, and P. Sicoli. This is the source of the translation of quotes used in this chapter.

8   Tim also recalls during this period that he got cluster headaches. These almost indescribably intense periods of pain would prevent him from sleeping during the day; by the end of an observing run, he was so sleep-deprived that he had hallucinations.

9   Tom Gehrels was born in the Netherlands and spent his teenage years fighting the Nazis as part of the Dutch Resistance. Much later, as an American astronomer, he viewed hazardous asteroids this way: "It's a well-known fact that when you have trouble within a country, the best way to divert attention from that is to find some

outside enemy. Here it would have the terrific effect of getting nations together, particularly the Russians and the United States. I have a hunch this would be a turning point in human history."

10   Initially, it was called WISE, which stands for Wide-field Infrared Survey Explorer. The asteroid-hunting part of the mission is called Near-Earth Object WISE, or NEOWISE. WISE was renamed NEOWISE when it started hunting asteroids full-time.

11   Artist and amateur astronomer Judy Schmidt processes NEOWISE images, with spectacular results. You can see her work at https://www.flickr.com/photos/geckzilla/.

12   I hope the next twenty years are so full of progress that this book seems equally dated in 2037.

13   That chapter is titled "The Lesson of Grand Forks: Can a Defense Against Asteroids Be Sustained?"

14   If you simply divide the volume of Jupiter by the volume of Earth, you get more than one thousand. But when I imagine this, I think of the Earth-sized planets like a pile of marbles, with gaps between them. Marbles randomly poured in a jar take up about 60 percent of the space in the jar. If you were to fill a Jupiter-sized sphere with Earth-sized planets in this random, non-orderly way, you'd be able to fit about eight hundred Earth-sized planets inside.

15   M&Ms are generally thirteen millimeters across. I know this because I bought a bag and measured them. Sometimes science is hard work.

16   This "notification plan" was followed in the case of 2008 TC_3. President George W. Bush's press secretary, Dana Perino, noted that the most unusual e-mail she received during her tenure was one titled "HEADS UP" at nine thirty p.m. informing her of the relatively small incoming asteroid and requesting that the Sudanese be notified through government channels.

FURTHER READING

A Wild Frontier

Chang, Kenneth. "You Could Actually Snooze Your Way Through an Asteroid Belt." *New York Times*, April 4, 2016.

Gladman, Brett et al. "The Resonant Trans-Neptunian Populations." *The Astronomical Journal* 144, no. 1 (June 2012).

Huang, Jiangchuan et al. "The Ginger-shaped Asteroid 4179 Toutatis: New Observations from a Successful Flyby of *Chang'e-2*." *Nature Scientific Reports* 3 (December 2013): doi:10.1038/srep03411.

Mainzer, Amy et al. "Characterizing Subpopulations within the Near-Earth Objects with *NEOWISE*: Preliminary Results." *The Astrophysical Journal* 752, no. 2 (June 2012).

Petit, J.-M. et al. "The Canada-France Ecliptic Plane Survey—Full Data Release: The Orbital Structure of the Kuiper Belt." *The Astronomical Journal* 142, no. 4 (September 2011).

Russell, C. T. et al. "Dawn at Vesta: Testing the Protoplanetary Paradigm." *Science* 336, no. 6082 (May 2012): doi:10.1126/science.1219381.

Things that Hit the Earth

Auer, Matthias and Mark K. Prior. "A New Era of Nuclear Test Verification." *Physics Today* 67, no. 9 (September 2014): doi:10.1063/PT.3.2516.

Barry, Ellen and Andrew E. Kramer. "Shock Wave of Fireball Meteor Rattles Siberia, Injuring 1,200." *New York Times*, February 15, 2013.

Brown, Peter G. et al. "A 500-kiloton Airburst over Chelyabinsk and an Enhanced Hazard from Small Impactors." *Nature* 503, no. 7475 (November 2013): 238–41. doi:10.1038/nature12741.

Ellington, M. J. "A Star Fell on Sylacauga: '54 Meteorite Struck Home, Woman, Changed Lives." *The Decatur Daily News*, November 30, 2006.

Kramer, Andrew E. "After Assault From the Heavens, Russians Search for Clues and Count Blessings." *New York Times*, February 16, 2013.

Kring, D. A. and M. Boslough. "Chelyabinsk: Portrait of an Asteroid Airburst." *Physics Today* 67, no. 9 (September 2014): 32–37. doi:10.1063/PT.3.2515.

Schulte, Peter et al. "The Chicxulub Asteroid Impact and Mass Extinction at the Cretaceous-Paleogene Boundary." *Science* 327, no. 5970 (March 2010): 1214–18. doi:10.1126/science.1177265.

Swindel, G. W. and W. B. Jones. "The Sylacauga, Talladega County, Alabama, Aerolite." *Meteoritics* 1, no. 2 (1954): 125–32.

Rules of Asteroid Hunting

Jenniskens, Peter and M. H. Shaddad. "2008 TC₃: The small asteroid with an impact." *Meteoritics & Planetary Science* 45, nos. 10–11 (2010): 1553–56. doi:10.1111/j.1945-5100.2010.01156.x.

Jenniskens, Peter et al. "The Impact and Recovery of Asteroid 2008 TC₃." *Nature* 458, no. 7237 (April 2009): 485–88. doi:10.1038/nature07920.

Kwok, Roberta. "The Rock that Fell to Earth." *Nature* 458 (March 2009): 401–3. doi:10.1038/458401a.

The First Asteroid

Heideman, Michael T., Don H. Johnson, and C. Sidney Burrus. "Gauss and the History of the Fast Fourier Transform." *IEEE ASSP Magazine* 1, no. 4 (October 1984): 14–21.

Hoskin, M. "Bode's Law and the Discovery of Ceres." In *Physics of Solar and Stellar Coronae: G. S. Vaiana Memorial Symposium*, edited by Jeffrey Linsky and Salvatore Serio. Dordrecht, Netherlands: Kluwer Academic Publishers, 1993, 35–46.

Serio, G. F., A. Manara, and P. Sicoli. "Giuseppe Piazzi and the Discovery of Ceres." In *Asteroids III*, edited by William Bottke Jr., Alberto Cellino, Paolo Paolicchi, and Richard P. Binzel. Tucson: University of Arizona Press, 2002, 17–24.

Terrestrial Asteroid Hunting

Bassett, Carol Ann. "If An Asteroid Threatens Earth, Tom Gehrels Will Sound the Alarm." *People* 22, no. 21 (November 1984).

Helin, Eleanor. "Identifying 'Interesting Objects' in Space." Web of Stories. http://www.webofstories.com/play/eleanor.helin/5;jsessionid=817D22F7532EF9DAC6D21E6E3F5B33AF.

Hughes, David W. and Brian G. Marsden. "Planet, Asteroid, Minor Planet: A Case Study in Astronomical Nomenclature." *Journal of Astronomical History and Heritage* 10, no. 1 (2007): 21–30.

The Giggle Factor

Gehrels, Tom, ed. *Hazards Due to Comets and Asteroids*. Tucson: University of Arizona Press, 1994.

Hotz, Robert Lee. "Jupiter Takes Huge Blow From Comet." *Los Angeles Times*, July 19, 1994. http://articles.latimes.com/1994-07-19/news/mn-17519_1_nuclear-weapons.

Howell, Elizabeth. "Shoemaker-Levy 9: Comet's Impact Left Its Mark on Jupiter." Space.com, February 19, 2013. http://www.space.com/19855-shoemaker-levy-9.html.

Ingersoll, A. P. and H. Kanamori. "Waves from the Collisions of Comet Shoemaker-Levy 9 with Jupiter." *Nature* 374, no. 6524 (April 1995): 706–8.

Ropeik, David. "When Past Disasters Are Prologue." *Nautilus*, no. 4 (August 8, 2013).

Takata, Toshiko et al. "Comet Shoemaker-Levy 9: Impact on Jupiter and Plume Evolution." *Icarus* 109, no. 1 (May 1994): 3–19. doi:10.1006/icar.1994.1074.

Weaver, H. A. et al. "Hubble Space Telescope Observations of Comet P/Shoemaker-Levy 9." *Science* 263, no. 5148 (February 1994): 787–91.

The Planetary Defense Coordination Office

Goldberg, Bruce. "Ex-Bush Press Secretary Perino Offers Peek into White House." *Denver Business Journal*, October 16, 2009. http://www.bizjournals.com/denver/stories/2009/10/12/daily88.html.

Planetary Defense Coordination Office website. https://www.nasa.gov/planetarydefense/overview.

ABOUT THE AUTHOR

Dr. Carrie Nugent is an asteroid hunter who works with a small team to discover and study asteroids at Caltech/IPAC. Dr. Nugent earned her PhD in Geophysics and Space Physics from UCLA in 2013 and is a 2016 TED Fellow. Specializing in thermophysical modeling, Dr. Nugent uses observations from the space-based infrared telescope NEOWISE to better understand asteroid surfaces. Asteroid 8801 Nugent was named in her honor.

In her free time, Dr. Nugent hosts and produces *Spacepod* (listentospacepod. com). On this weekly podcast, she invites astronomers, planetary scientists, and engineers to sit, share a drink, and tell the world about their corner of the cosmos.

WATCH CARRIE NUGENT'S TED TALK

Carrie's TED Talk, available for free at TED.com, is the companion to *Asteroid Hunters*.

PHOTO: RYAN LASH / TED

Jedidah Isler
How I Fell in Love with Quasars, Blazars, and Our Incredible Universe
Jedidah Isler first fell in love with the night sky as a little girl. Now she's an astrophysicist who studies supermassive hyperactive black holes. In a charming talk, she takes us trillions of kilometers from Earth to introduce us to objects that can be one to ten billion times the mass of the sun — and that shoot powerful jet streams of particles in our direction.

Sarah Parcak
Hunting for Peru's Lost Civilizations—with Satellites
Around the world, hundreds of thousands of lost ancient sites lie buried and hidden from view. Satellite archaeologist Sarah Parcak is determined to find them before looters do. With the 2016 TED Prize, Parcak is building an online citizen-science tool called GlobalXplorer that will train an army of volunteer explorers to find and protect the world's hidden heritage. In this talk, she offers a preview of the first place they'll look: Peru — the home of Machu Picchu, the Nazca lines, and other archaeological wonders waiting to be discovered.

Allan Adams
What the Discovery of Gravitational Waves Means
More than a billion years ago, two black holes in a distant galaxy locked into a spiral, falling inexorably toward each other and colliding. "All that energy was pumped into the fabric of time and space itself," says theoretical physicist Allan Adams, "making the universe explode in roiling waves of gravity." About twenty-five years ago, a group of scientists built a giant laser detector called LIGO to search for these kinds of waves, which had been predicted but never observed. In this mind-bending talk, Adams breaks down what happened when, in September 2015, LIGO detected an unthinkably small anomaly, leading to one of the most exciting discoveries in the history of physics.

Lucianne Walkowicz
Finding Planets around Other Stars
How do we find planets — even habitable planets — around other stars? By looking for tiny dimming as a planet passes in front of its sun, TED Fellow Lucianne Walkowicz and the Kepler mission have found some 1,200 potential new planetary systems. With new techniques, they may even find ones with the right conditions for life.

The Misfit's Manifesto
by Lidia Yuknavitch

Misfit: A person who missed fitting in, a person who fits in badly. It's a word no one typically tries to own. Until now. Bestselling author Lidia Yuknavitch reveals why she is proud to be a misfit.

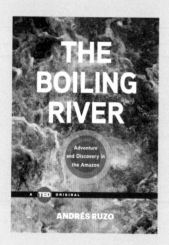

The Boiling River
Adventure and Discovery in the Amazon
by Andrés Ruzo

In this exciting adventure mixed with amazing scientific discovery, a young, exuberant explorer and geoscientist journeys deep into the Amazon—where rivers boil and legends come to life.

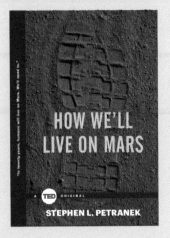

How We'll Live on Mars
by Stephen Petranek

Within twenty years, humans will live on Mars. We'll need to. Award-winning journalist Stephen Petranek makes the case that living on Mars is not just plausible but inevitable, thanks to new technology and the competitive spirit of the world's most forward-looking entrepreneurs.

The Great Questions of Tomorrow
by David Rothkopf

We stand on the cusp of a sweeping revolution—one that will impact every facet of our lives, from money to war to government. In this pivotal moment, David Rothkopf asks, what are the great questions we need to ask to best navigate our way forward?